Space, Time and Matter

Ashay Dharwadker

Vinay Dharwadker

INSTITUTE OF MATHEMATICS
H-501 PALAM VIHAR, GURGAON, HARYANA 122017, INDIA
www.dharwadker.org

Abstract

We show how the grand unified theory based on the proof of the four color theorem can be obtained entirely in terms of the Poincaré group of isometries of space and time. Electric and gauge charges of all the particles of the standard model can now be interpreted as elements of the Poincaré group. We define the space and time chiralities of all spin 1/2 fermions in agreement with Dirac's relativistic wave equation. All the particles of the standard model now correspond to irreducible representations of the Poincaré group according to Wigner's classification. Finally, we construct the Steiner system of fermions and show how the Mathieu group acts as the group of symmetries of the fundamental building blocks of matter.

Contents

Introduction

We show how the grand unified theory based on the proof of the four color theorem **[1][2][3]**, can be obtained entirely in terms of the Poincaré group of isometries of space and time. The grand unified theory is formulated in terms of the split extension $\boldsymbol{Z}_4]S_3$ of the cyclic group $\boldsymbol{Z}_4 = \{\underline{0}, \underline{1}, \underline{2}, \underline{3}\}$ of integers with addition modulo 4 by the symmetric group on three letters $S_3 = \{1, \rho, \rho^2, \sigma, \sigma\rho, \sigma\rho^2\}$. All particles of the standard model are represented by Schrödinger discs on the particle frame which forms the gauge at each point of space-time. The four electric charges 0, 1/3, 2/3, 1 are represented by $\underline{0}, \underline{1}, \underline{2}, \underline{3}$; the single electromagnetic gauge charge is represented by 1; the two weak gauge charges are represented by ρ, ρ^2; the three strong gauge charges are represented by $\sigma, \sigma\rho, \sigma\rho^2$; the five gravitational gauge charges are represented by $\underline{0}, \underline{1}, \underline{2}, \underline{3}$ and σ. The gauge charges form the parameters of the electromagnetic gauge group $\boldsymbol{U}(1)$, the weak gauge group $\boldsymbol{SU}(2)$, the strong gauge group $\boldsymbol{SU}(3)$ and the gravitational gauge group $\boldsymbol{SU}(5)$ respectively. The electric and gauge charges obtain $\pm$ signs from scalar multiplication in the group algebra of the split extension $\boldsymbol{Z}_4]S_3$. All fermions and bosons of the standard model obtain signed electric and gauge charges via the labeling of the particle frame by elements of the group algebra of the split extension $\boldsymbol{Z}_4]S_3$. The gauge groups are embedded in a sequence $\boldsymbol{U}(1) \to \boldsymbol{SU}(2) \to \boldsymbol{SU}(3) \to \boldsymbol{SU}(5)$ in the grand unification.

In Section 1 on **ISOMETRIES** we define the Poincaré group and show how it forms a split extension of Minkowski space-time by the Lorentz group. In Section 2 on **CHARGES** we show how the electric charges $\underline{0}, \underline{1}, \underline{2}, \underline{3}$ are obtained in terms of time translations in the Poincaré group. This corresponds to Dirac's abstract construction of the magnetic monopole, however magnetic monopoles cannot be observable particles in the standard model. Comparison with 't Hooft's model shows that this construction also explains quark confinement. We show that the electroweak and strong gauge charges $1, \rho, \rho^2, \sigma, \sigma\rho, \sigma\rho^2$ are Lorentz transformations. Thus, all electric and gauge charges are obtained as elements of the Poincaré group. In Section 3 on **CHIRALITIES** we show each spin 1/2 fermion is obtained as a solution of Dirac's wave equation on the particle frame. We obtain the traditional definition of chirality on the particle frame in terms of the fifth gamma matrix γ^5 in the Weyl basis of the Clifford algebra of space-time. However, Dirac's definition of chirality is only part of the picture; we show that it is natural to define Dirac's chirality as the time-chirality χ_T of a fermion and there is a dual notion of space-chirality χ_S associated with each fermion. Now we have the complete picture: the time-chirality χ_T distinguishes particles from antiparticles and the space-chirality χ_S distinguishes quarks from leptons in the standard model. In Section 4 on **REPRESENTATIONS** we show how all the particles of the standard model correspond to irreducible representations of the Poincaré group according to Wigner's classification. Finally, we construct the Steiner system of fermions $S(5, 8, 24)$ according to the proof of the four color theorem and show how the Mathieu group $\mathbf{M}_{24}$ acts as the group of symmetries of the fundamental building blocks of matter.

1. Isometries

The space-time of physics is defined by four real coordinates: the three space coordinates X, Y, Z and the time coordinate T. The theory of special relativity **[4][5]** is concerned with inertial reference frames in which force-free particles do not experience any acceleration with respect to the coordinate system. Inertial reference frames are defined by the group of Lorentz transformations which are linear transformations of the space-time coordinates that leave the velocity of light c invariant. A Lorentz transformation transforms one inertial reference frame to another that is in uniform motion relative to the first. One of the main motivations for restricting the theory to inertial reference frames is that Maxwell's equations for electromagnetism and the Yang-Mills equations for the weak and strong fields remain unchanged if the space-time coordinates are subjected to Lorentz transformations. Thus, according to the theory of special relativity, light has a constant velocity of propagation c. If a light signal in a vacuum starts from a space point (X, Y, Z) at the time T, it spreads as a spherical wave and reaches a neighboring space point $(X+dX, Y+dY, Z+dZ)$ at the time $T+dT$. Measuring the distance traveled by the light signal, we must have

$$(cdT)^2 = (dX)^2+(dY)^2+(dZ)^2 \tag{1.1}$$

Figure 1.1. *A light signal is represented by a sphere of radius cdT centered at (X, Y, Z)*

The equation (1.1) may be rewritten as

$$(dX)^2 + (dY)^2 + (dZ)^2 - (cdT)^2 = 0 \tag{1.2}$$

Equation (1.2) represents an objective relation between neighboring space-time points and it holds for all inertial reference frames provided the transformations of the coordinates are restricted to those of special relativity, i.e. Lorentz transformations. By considering the inertial reference frames of special relativity, it can also be shown that the Lorentz transformations are precisely the linear transformations that leave the more general quantity

$$(dS)^2 = (dX)^2 + (dY)^2 + (dZ)^2 - (cdT)^2 \tag{1.3}$$

invariant. Note, however, that the vanishing of $(dS)^2$ in equation (1.3) does not imply that the two space-time points coincide; it means that the two space-time points can be connected by a light signal. This is Einstein's physical motivation for the theory of special relativity.

Let us now formulate the theory of special relativity in Minkowski's notation **[6]**. Select fundamental Planck units **[7][8]** for measuring X, Y, Z and T, so that the velocity of light $c = 1$. Then *Minkowski space-time* is written as $\mathbf{R}^{(3,\,1)} = \{(X, Y, Z, T) \mid X, Y, Z, T \in \mathbf{R}\}$, which forms a 4-dimensional vector space over the real numbers $\mathbf{R}$ with the usual addition of vectors and scalar multiplication. The *Lorentz inner product* in $\mathbf{R}^{(3,\,1)}$ is defined as

$$(X_1, Y_1, Z_1, T_1) \cdot (X_2, Y_2, Z_2, T_2) = X_1X_2 + Y_1Y_2 + Z_1Z_2 - T_1T_2 \tag{1.4}$$

and the *Lorentz norm* in $\mathbf{R}^{(3,\,1)}$ is defined as

$$\begin{aligned} |(X, Y, Z, T)| &= ((X, Y, Z, T) \cdot (X, Y, Z, T))^{1/2} \\ &= (X^2 + Y^2 + Z^2 - T^2)^{1/2} \end{aligned} \tag{1.5}$$

which is a complex number in general. The *Lorentz metric* in $\mathbf{R}^{(3,\,1)}$ is given by

$$\begin{aligned} \eta((X_1, Y_1, Z_1, T_1), (X_2, Y_2, Z_2, T_2)) &= |(X_1, Y_1, Z_1, T_1) - (X_2, Y_2, Z_2, T_2)| \\ &= |(X_1 - X_2, Y_1 - Y_2, Z_1 - Z_2, T_1 - T_2)| \\ &= ((X_1 - X_2)^2 + (Y_1 - Y_2)^2 + (Z_1 - Z_2)^2 - (T_1 - T_2)^2)^{1/2} \end{aligned} \tag{1.6}$$

in agreement with (1.3) which is an expression of the Lorentz metric locally in terms of infinitesimal differentials. A linear transformation $f : \mathbf{R}^{(3,\,1)} \rightarrow \mathbf{R}^{(3,\,1)}$ is called a *Lorentz transformation* if

$$f(X_1, Y_1, Z_1, T_1) \cdot f(X_2, Y_2, Z_2, T_2) = (X_1, Y_1, Z_1, T_1) \cdot (X_2, Y_2, Z_2, T_2) \tag{1.7}$$

for all (X_1, Y_1, Z_1, T_1) and (X_2, Y_2, Z_2, T_2) in $\mathbf{R}^{(3,\,1)}$. The Lorentz transformations form a group $\mathbf{O}^{(3,\,1)}$ called the *Lorentz group* under the binary operation of function composition. The Lorentz group $\mathbf{O}^{(3,\,1)}$ consists precisely of all the transformations of Minkowski space-time $\mathbf{R}^{(3,\,1)}$ that leave the velocity of light $c = 1$ invariant.

Referring to Lemma 4 **[1][3]**, let $S_3 = \{1, \rho, \rho^2, \sigma, \sigma\rho, \sigma\rho^2\}$ denote the symmetric group on three letters which is abstractly isomorphic to the dihedral group of order 6 generated by σ, ρ subject to the relations $\sigma^2 = 1$, $\rho^3 = 1$ and $\sigma\rho\sigma^{-1} = \rho^{-1}$. We identify S_3 with transformations of Minkowski space-time $\mathbf{R}^{(3,\ 1)}$ via the following correspondence:

$$1 = \begin{bmatrix} 1 & 0 & 0 & 0 \\ 0 & 1 & 0 & 0 \\ 0 & 0 & 1 & 0 \\ 0 & 0 & 0 & 1 \end{bmatrix} \quad \rho = \begin{bmatrix} 0 & 1 & 0 & 0 \\ 0 & 0 & 1 & 0 \\ 1 & 0 & 0 & 0 \\ 0 & 0 & 0 & 1 \end{bmatrix} \quad \rho^2 = \begin{bmatrix} 0 & 0 & 1 & 0 \\ 1 & 0 & 0 & 0 \\ 0 & 1 & 0 & 0 \\ 0 & 0 & 0 & 1 \end{bmatrix}$$

$$\sigma = \begin{bmatrix} 0 & 1 & 0 & 0 \\ 1 & 0 & 0 & 0 \\ 0 & 0 & 1 & 0 \\ 0 & 0 & 0 & 1 \end{bmatrix} \quad \sigma\rho = \begin{bmatrix} 0 & 0 & 1 & 0 \\ 0 & 1 & 0 & 0 \\ 1 & 0 & 0 & 0 \\ 0 & 0 & 0 & 1 \end{bmatrix} \quad \sigma\rho^2 = \begin{bmatrix} 1 & 0 & 0 & 0 \\ 0 & 0 & 1 & 0 \\ 0 & 1 & 0 & 0 \\ 0 & 0 & 0 & 1 \end{bmatrix} \tag{1.8}$$

It is easy to verify that the six matrices (1.8) satisfy the generating relations of S_3 under matrix multiplication and form an isomorphic group of transformations S_3 of Minkowski space-time $\mathbf{R}^{(3,\ 1)}$. The transformation group S_3 is the permutation group on the space coordinates X, Y, Z that keeps the time coordinate T fixed. Acting on a right-handed space coordinate system, the transformations 1, ρ, ρ^2 give right-handed space coordinate systems whereas the transformations σ, $\sigma\rho$, $\sigma\rho^2$ give left-handed space coordinate systems as shown below in (1.9):

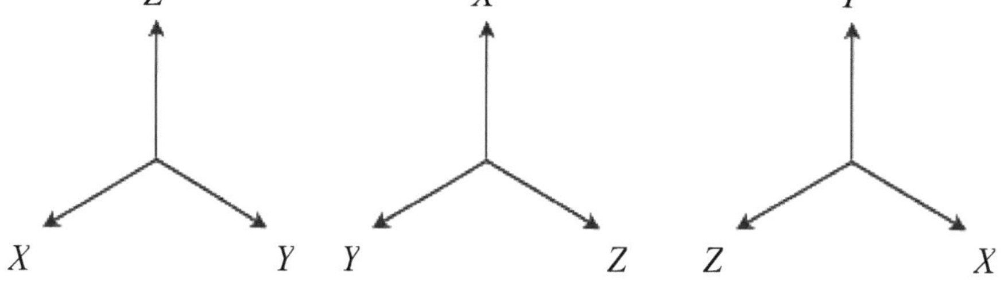

(1.9)

$$1 \quad \begin{bmatrix} X \\ Y \\ Z \\ T \end{bmatrix} = \begin{bmatrix} X \\ Y \\ Z \\ T \end{bmatrix} \qquad \rho \quad \begin{bmatrix} X \\ Y \\ Z \\ T \end{bmatrix} = \begin{bmatrix} Y \\ Z \\ X \\ T \end{bmatrix} \qquad \rho^2 \quad \begin{bmatrix} X \\ Y \\ Z \\ T \end{bmatrix} = \begin{bmatrix} Z \\ X \\ Y \\ T \end{bmatrix}$$

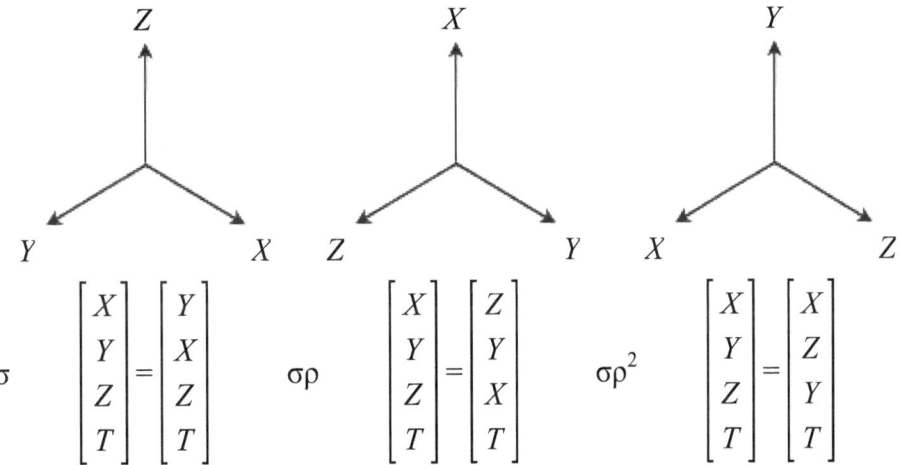

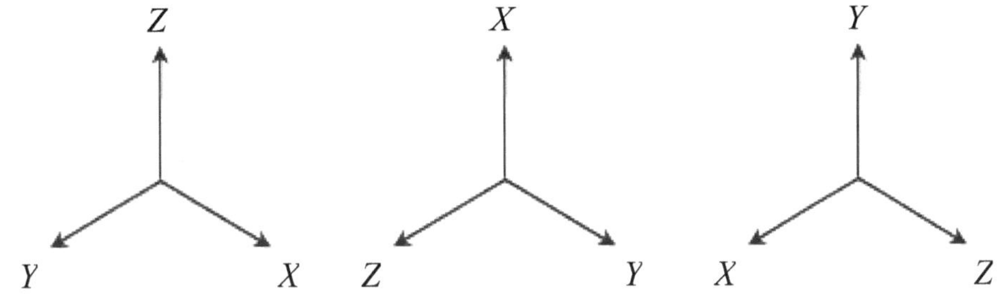

$$\sigma \begin{bmatrix} X \\ Y \\ Z \\ T \end{bmatrix} = \begin{bmatrix} Y \\ X \\ Z \\ T \end{bmatrix} \qquad \sigma\rho \begin{bmatrix} X \\ Y \\ Z \\ T \end{bmatrix} = \begin{bmatrix} Z \\ Y \\ X \\ T \end{bmatrix} \qquad \sigma\rho^2 \begin{bmatrix} X \\ Y \\ Z \\ T \end{bmatrix} = \begin{bmatrix} X \\ Z \\ Y \\ T \end{bmatrix}$$

By (1.8) and (1.9), each transformation in S_3 leaves the expression (1.4) for the Lorentz inner product invariant and hence satisfies the condition (1.7) for being a Lorentz transformation. Thus, S_3 is a subgroup of the Lorentz group $\mathbf{O}^{(3,\,1)}$.

A linear transformation $f: \mathbf{R}^{(3,\,1)} \to \mathbf{R}^{(3,\,1)}$ is called an *isometry* if

$$\eta(f(X_1, Y_1, Z_1, T_1), f(X_2, Y_2, Z_2, T_2)) = \eta((X_1, Y_1, Z_1, T_1), (X_2, Y_2, Z_2, T_2)) \qquad (1.10)$$

for all (X_1, Y_1, Z_1, T_1) and (X_2, Y_2, Z_2, T_2) in $\mathbf{R}^{(3,\,1)}$. The isometries form a group $\mathbf{I}^{(3,\,1)}$ called the *Poincaré group* under the binary operation of function composition. The Poincaré group $\mathbf{I}^{(3,\,1)}$ consists precisely of all the transformations of Minkowski space-time $\mathbf{R}^{(3,\,1)}$ that leave the distance between every pair of space-time points in the metric η invariant. If f is a Lorentz transformation in $\mathbf{O}^{(3,\,1)}$, then

$$\begin{aligned}
& \eta(f(X_1, Y_1, Z_1, T_1), f(X_2, Y_2, Z_2, T_2)) \\
= \; & |f(X_1, Y_1, Z_1, T_1) - f(X_2, Y_2, Z_2, T_2)| && \text{(by 1.6)} \\
= \; & |f((X_1, Y_1, Z_1, T_1) - (X_2, Y_2, Z_2, T_2))| && \text{(by linearity of } f\text{)} \\
= \; & |f(X_1 - X_2, Y_1 - Y_2, Z_1 - Z_2, T_1 - T_2)| && \text{(using vector space } \mathbf{R}^{(3,\,1)}\text{)} \\
= \; & (f(X_1 - X_2, Y_1 - Y_2, Z_1 - Z_2, T_1 - T_2) \cdot f(X_1 - X_2, Y_1 - Y_2, Z_1 - Z_2, T_1 - T_2))^{1/2} && \text{(by 1.5)} \qquad (1.11) \\
= \; & ((X_1 - X_2, Y_1 - Y_2, Z_1 - Z_2, T_1 - T_2) \cdot (X_1 - X_2, Y_1 - Y_2, Z_1 - Z_2, T_1 - T_2))^{1/2} && \text{(by 1.7, since } f \text{ is in } \mathbf{O}^{(3,\,1)}\text{)} \\
= \; & |(X_1 - X_2, Y_1 - Y_2, Z_1 - Z_2, T_1 - T_2)| && \text{(by 1.5)} \\
= \; & |(X_1, Y_1, Z_1, T_1) - (X_2, Y_2, Z_2, T_2)| && \text{(using vector space } \mathbf{R}^{(3,\,1)}\text{)} \\
= \; & \eta((X_1, Y_1, Z_1, T_1), (X_2, Y_2, Z_2, T_2)) && \text{(by 1.6)}
\end{aligned}$$

for all (X_1, Y_1, Z_1, T_1) and (X_2, Y_2, Z_2, T_2) in $\mathbf{R}^{(3,\,1)}$, showing that the Lorentz group $\mathbf{O}^{(3,\,1)}$ is a subgroup of the Poincaré group $\mathbf{I}^{(3,\,1)}$.

Given a fixed (X', Y', Z', T') in $\mathbf{R}^{(3,1)}$, a *translation* of Minkowski space-time by (X', Y', Z', T') is a linear transformation $f_{(X', Y', Z', T')} : \mathbf{R}^{(3,1)} \to \mathbf{R}^{(3,1)}$ given by

$$f_{(X', Y', Z', T')}(X, Y, Z, T) = (X + X', Y + Y', Z + Z', T + T') \tag{1.12}$$

for all (X, Y, Z, T) in $\mathbf{R}^{(3,1)}$. Let $\mathbf{R}'^{(3,1)}$ denote the set of all translations of Minkowski space-time. Given two translations $f_{(X', Y', Z', T')}$ and $f_{(X'', Y'', Z'', T'')}$, the binary operation of function composition $f_{(X', Y', Z', T')}f_{(X'', Y'', Z'', T'')}$ is well-defined:

$$\begin{aligned} & f_{(X', Y', Z', T')}f_{(X'', Y'', Z'', T'')}(X, Y, Z, T) \\ = \ & f_{(X', Y', Z', T')}(X + X'', Y + Y'', Z + Z'', T + T'') \\ = \ & (X + X' + X'', Y + Y' + Y'', Z + Z' + Z'', T + T' + T'') \\ = \ & f_{(X' + X'', Y' + Y'', Z' + Z'', T' + T'')} \end{aligned} \tag{1.13}$$

Under the correspondence $f_{(X', Y', Z', T')} \to (X', Y', Z', T')$, the translations $\mathbf{R}'^{(3,1)}$ of Minkowski space-time form an abelian group isomorphic to the additive group of the vector space of Minkowski space-time $\mathbf{R}^{(3,1)}$.

We shall now show that the only translation that is also a Lorentz transformation is the identity transformation of Minkowski space-time, i.e. $\mathbf{R}'^{(3,1)} \cap \mathbf{O}^{(3,1)} = \{1\}$. Suppose $f_{(X', Y', Z', T')} = g$, where $f_{(X', Y', Z', T')}$ is a translation in $\mathbf{R}'^{(3,1)}$ and g is a Lorentz transformation in $\mathbf{O}^{(3,1)}$. Then, for all (X, Y, Z, T) in $\mathbf{R}^{(3,1)}$:

$$\begin{aligned} g(X, Y, Z, T) &= f_{(X', Y', Z', T')}(X, Y, Z, T) \\ &= (X + X', Y + Y', Z + Z', T + T') \end{aligned} \tag{1.14}$$

In particular,

$$\begin{aligned} g(0, 0, 0, 0) &= f_{(X', Y', Z', T')}(0, 0, 0, 0) \\ &= (X', Y', Z', T') \end{aligned} \tag{1.15}$$

Then, since g is a Lorentz transformation, for all (X, Y, Z, T) in $\mathbf{R}^{(3,1)}$:

$$\begin{aligned} 0 &= X \cdot 0 + Y \cdot 0 + Z \cdot 0 - T \cdot 0 \\ &= (X, Y, Z, T) \cdot (0, 0, 0, 0) && \text{(by 1.4)} \\ &= g(X, Y, Z, T) \cdot g(0, 0, 0, 0) && \text{(by 1.7)} \\ &= (X + X', Y + Y', Z + Z', T + T') \cdot (X', Y', Z', T') && \text{(by 1.14 and 1.15)} \end{aligned} \tag{1.16}$$

$$= (X + X')X' + (Y + Y')Y' + (Z + Z')Z' - (T + T')T' \quad \text{(by 1.4)}$$

$$= (XX' + YY' + ZZ' - TT') + (X'^2 + Y'^2 + Z'^2 - T'^2)$$

Put $T = 0$ in (1.16), then for all X, Y, Z:

$$T'^2 = (XX' + YY' + ZZ') + (X'^2 + Y'^2 + Z'^2) \quad (1.17)$$

The LHS of (1.17) is non-negative, but we can always choose X, Y, Z such that the RHS of (1.17) is negative unless $X' = Y' = Z' = 0$. Thus $f_{(X', Y', Z', T')} = g$ must be the identity transformation of Minkowski space-time. This implies that $\mathbf{R'}^{(3,1)} \cap \mathbf{O}^{(3,1)} = \{1\}$.

We shall now show that the only isometries that fix the origin of Minkowski space-time are the Lorentz transformations. Suppose f is an isometry of Minkowski space-time such that $f(0, 0, 0, 0) = (0, 0, 0, 0)$. Given any (X_1, Y_1, Z_1, T_1) and (X_2, Y_2, Z_2, T_2) in $\mathbf{R}^{(3,1)}$, define

$$\begin{aligned} (X_1^*, Y_1^*, Z_1^*, T_1^*) &= f(X_1, Y_1, Z_1, T_1) \\ (X_2^*, Y_2^*, Z_2^*, T_2^*) &= f(X_2, Y_2, Z_2, T_2) \end{aligned} \quad (1.18)$$

Then

$$\begin{aligned} & X_1^{\,2} + Y_1^2 + Z_1^{\,2} - T_1^2 \\ = \; & \eta((X_1, Y_1, Z_1, T_1), (0, 0, 0, 0))^2 && \text{(by 1.6)} \\ = \; & \eta(f(X_1, Y_1, Z_1, T_1), f(0, 0, 0, 0))^2 && \text{(since } f \text{ is an isometry)} \quad (1.19) \\ = \; & \eta((X_1^*, Y_1^*, Z_1^*, T_1^*), (0, 0, 0, 0))^2 && \text{(by 1.18)} \\ = \; & X_1^{*\,2} + Y_1^{*2} + Z_1^{*\,2} - T_1^{*2} && \text{(by 1.6)} \end{aligned}$$

Similarly

$$\begin{aligned} & X_2^{\,2} + Y_2^2 + Z_2^{\,2} - T_2^2 \\ = \; & \eta((X_2, Y_2, Z_2, T_2), (0, 0, 0, 0))^2 && \text{(by 1.6)} \\ = \; & \eta(f(X_2, Y_2, Z_2, T_2), f(0, 0, 0, 0))^2 && \text{(since } f \text{ is an isometry)} \quad (1.20) \\ = \; & \eta((X_2^*, Y_2^*, Z_2^*, T_2^*), (0, 0, 0, 0))^2 && \text{(by 1.18)} \\ = \; & X_2^{*\,2} + Y_2^{*2} + Z_2^{*\,2} - T_2^{*2} && \text{(by 1.6)} \end{aligned}$$

Now, since f is an isometry

$$\begin{aligned}
& \eta(f(X_1, Y_1, Z_1, T_1), f(X_2, Y_2, Z_2, T_2))^2 = \eta((X_1, Y_1, Z_1, T_1), (X_2, Y_2, Z_2, T_2))^2 && \text{(by 1. 0)} \\
\Rightarrow\ & \eta((X_1^*, Y_1^*, Z_1^*, T_1^*), (X_2^*, Y_2^*, Z_2^*, T_2^*))^2 = \eta((X_1, Y_1, Z_1, T_1), (X_2, Y_2, Z_2, T_2))^2 && \text{(by 1.18)} \\
\Rightarrow\ & (X_1^* - X_2^*)^2 + (Y_1^* - Y_2^*)^2 + (Z_1^* - Z_2^*)^2 - (T_1^* - T_2^*)^2 = (X_1 - X_2)^2 + (Y_1 - Y_2)^2 + (Z_1 - Z_2)^2 - (T_1 - T_2)^2 && \text{(by 1.6)}
\end{aligned} \tag{1.21}$$

Expanding both sides of the last equation in (1.21) we obtain

$$A^* + B^* + C^* = A + B + C \tag{1.22}$$

where

$$\begin{aligned}
A^* &= X_1^{*\,2} + Y_1^{*2} + Z_1^{*\,2} - T_1^{*2} & \qquad A &= X_1^{\,2} + Y_1^2 + Z_1^{\,2} - T_1^2 \\
B^* &= X_2^{*\,2} + Y_2^{*2} + Z_2^{*\,2} - T_2^{*2} & \qquad B &= X_2^{\,2} + Y_2^2 + Z_2^{\,2} - T_2^2 \\
C^* &= -2\,X_1^*X_2^* -2\,Y_1^*Y_2^* -2\,Z_1^*Z_2^* +2\,T_1^*T_2^* & \qquad C &= -2\,X_1X_2 -2\,Y_1Y_2 -2\,Z_1Z_2 +2\,T_1T_2
\end{aligned}$$

In equation (1.22), summands A^*, A are equal by (1.19) and summands B^*, B are equal by (1.20). Hence summands C^*, C must be equal:

$$\begin{aligned}
& -2\,X_1^*X_2^* -2\,Y_1^*Y_2^* -2\,Z_1^*Z_2^* +2\,T_1^*T_2^* = -2\,X_1X_2 -2\,Y_1Y_2 -2\,Z_1Z_2 +2\,T_1T_2 \\
\Rightarrow\ & X_1^*X_2^* + Y_1^*Y_2^* + Z_1^*Z_2^* - T_1^*T_2^* = X_1X_2 + Y_1Y_2 + Z_1Z_2 - T_1T_2 && \text{(dividing by -2)} \\
\Rightarrow\ & (X_1^*, Y_1^*, Z_1^*, T_1^*) \cdot (X_2^*, Y_2^*, Z_2^*, T_2^*) = (X_1, Y_1, Z_1, T_1) \cdot (X_2, Y_2, Z_2, T_2) && \text{(by 1.4)} \\
\Rightarrow\ & f(X_1, Y_1, Z_1, T_1) \cdot f(X_2, Y_2, Z_2, T_2) = (X_1, Y_1, Z_1, T_1) \cdot (X_2, Y_2, Z_2, T_2) && \text{(by 1.18)}
\end{aligned} \tag{1.23}$$

Hence, if f is an isometry of Minkowski space-time such that $f(0, 0, 0, 0) = (0, 0, 0, 0)$ then f must be a Lorentz transformation.

Using this result, we shall now show that every isometry of Minkowski space-time can be uniquely written as the product of a translation and a Lorentz transformation. Suppose f is an isometry of Minkowski space-time. To demonstrate the existence, we want to show that $f = f_{(X', Y', Z', T')}\,g$ for at least one translation $f_{(X', Y', Z', T')}$ in $\mathbf{R'}^{(3,\,1)}$ and at least one Lorentz transformation g in $\mathbf{O}^{(3,\,1)}$. Define $(X', Y', Z', T') = f(0, 0, 0, 0)$ and $g = f_{(X', Y', Z', T')}^{\;-1} f$. Then $f_{(X', Y', Z', T')}$ is a translation in $\mathbf{R'}^{(3,\,1)}$ and as a product of two isometries, g is certainly an isometry. We must show that g is a Lorentz transformation:

$$\begin{aligned}
g(0, 0, 0, 0) &= f_{(X', Y', Z', T')}^{\;-1} f(0, 0, 0, 0) \\
&= f_{(X', Y', Z', T')}^{\;-1} (X', Y', Z', T') \\
&= (0, 0, 0, 0)
\end{aligned} \tag{1.24}$$

Thus, g is a Lorentz transformation by (1.23). To demonstrate the uniqueness, suppose $f = f_{(X'', Y'', Z'', T'')}\,g'$ for some translation $f_{(X'', Y'', Z'', T'')}$ in $\mathbf{R'}^{(3,\,1)}$ and some Lorentz transformation g' in $\mathbf{O}^{(3,1)}$:

$$\Rightarrow \quad f_{(X', Y', Z', T')}\, g \quad = \quad f \quad = \quad f_{(X'', Y'', Z'', T'')}\, g' \tag{1.25}$$

$$\Rightarrow \quad f_{(X'', Y'', Z'', T'')}^{-1} f_{(X', Y', Z', T')} \quad = \quad g' g^{-1}$$

But $f_{(X'', Y'', Z'', T'')}^{-1} f_{(X', Y', Z', T')} = g'g^{-1}$ belongs to $\mathbf{R}'^{(3,1)} \cap \mathbf{O}^{(3,1)} = \{1\}$ by (1.17). Hence $f_{(X'', Y'', Z'', T'')}^{-1} f_{(X', Y', Z', T')} = g'g^{-1} = 1$. Thus $f_{(X'', Y'', Z'', T'')} = f_{(X', Y', Z', T')}$ and $g' = g$, demonstrating the uniqueness of the isometry $f = f_{(X', Y', Z', T')}\, g$ as the product of a translation and a Lorentz transformation.

We shall now show how the Lorentz group $\mathbf{O}^{(3,1)}$ acts as a group of automorphisms of the vector space of translations $\mathbf{R}'^{(3,1)}$ of Minkowski space-time. The abelian group (module) of the vector space $\mathbf{R}'^{(3,1)}$ is isomorphic to the abelian group (module) $\mathbf{R}^{(3,1)}$ of Minkowski space-time under the correspondence $f_{(X', Y', Z', T')} \rightarrow (X', Y', Z', T')$. Let $Aut\ \mathbf{R}'^{(3,1)}$ denote the group of automorphisms of the module $\mathbf{R}'^{(3,1)}$. Define the group representation

$$\begin{aligned} \xi: \quad \mathbf{O}^{(3,1)} &\rightarrow Aut\ \mathbf{R}'^{(3,1)} \\ g &\rightarrow \xi(g) \end{aligned} \tag{1.26}$$

where the right action of $\xi(g)$ on $\mathbf{R}'^{(3,1)}$ is given by

$$\begin{aligned} \xi(g): \quad \mathbf{R}'^{(3,1)} &\rightarrow \mathbf{R}'^{(3,1)} \\ f &\rightarrow g^{-1} f g \end{aligned} \tag{1.27}$$

Note that we write the action on the right as $f\,\xi(g) = g^{-1} f g$ for each f in $\mathbf{R}'^{(3,1)}$. To show that ξ is a group representation, we must verify that ξ is a homomorphism. For each f in $\mathbf{R}'^{(3,1)}$:

$$\begin{aligned} f\,\xi(g_1 g_2) &= (g_1 g_2)^{-1} f (g_1 g_2) \\ &= g_2^{-1}(g_1^{-1} f g_1) g_2 \\ &= g_2^{-1}(f\,\xi(g_1)) g_2 \\ &= (f\,\xi(g_1))\xi(g_2) \\ &= f\,\xi(g_1)\xi(g_2) \end{aligned} \tag{1.28}$$

Hence, ξ is a homomorphism showing how the Lorentz group $\mathbf{O}^{(3,1)}$ acts as a group of automorphisms of the vector space of translations $\mathbf{R}'^{(3,1)}$ of Minkowski space-time. Following **[1]** Section III, the abelian group of the vector space of translations $\mathbf{R}'^{(3,1)}$ is an Eilenberg module for the Lorentz group $\mathbf{O}^{(3,1)}$ and we can write the Poincaré group $\mathbf{I}^{(3,1)}$ of isometries of Minkowski space-time as the split extension $\mathbf{R}'^{(3,1)}]\mathbf{O}^{(3,1)}$:

$$\mathbf{I}^{(3,\,1)} \;=\; \mathbf{R'}^{(3,\,1)}]\mathbf{O}^{(3,\,1)} \;=\; \{(f, g) | f \in \mathbf{R'}^{(3,\,1)}, g \in \mathbf{O}^{(3,\,1)}\} \tag{1.29}$$

By (1.24) and (1.25), each element fg of the Poincaré group $\mathbf{I}^{(3,\,1)}$ is written uniquely as the element (f, g) of the split extension $\mathbf{R'}^{(3,\,1)}]\mathbf{O}^{(3,\,1)}$. The group operation in the split extension $\mathbf{R'}^{(3,\,1)}]\mathbf{O}^{(3,\,1)}$ is given by:

$$\begin{aligned}(f_1, g_1)(f_2, g_2) &= (f_1(f_2\,\xi(g_1)), g_1g_2) \\ &= (f_1(g_1^{-1}f_2g_1), g_1g_2)\end{aligned} \tag{1.30}$$

There is an exact sequence of groups

$$\mathbf{R'}^{(3,\,1)} \xrightarrow{\iota^{(3,\,1)}} \mathbf{I}^{(3,\,1)} = \mathbf{R'}^{(3,\,1)}]\mathbf{O}^{(3,\,1)} \xrightarrow{\pi^{(3,\,1)}} \mathbf{O}^{(3,\,1)} \tag{1.31}$$

where the injection function is defined by $\iota^{(3,\,1)}(f) = (f, 1)$, the projection function is defined by $\pi^{(3,\,1)}((f, g)) = g$ and the sequence is split by the zero function

$$\mathbf{I}^{(3,\,1)} = \mathbf{R'}^{(3,\,1)}]\mathbf{O}^{(3,\,1)} \xleftarrow{o^{(3,\,1)}} \mathbf{O}^{(3,\,1)} \tag{1.32}$$

defined by $o^{(3,\,1)}(g) = (f_{(0,\,0,\,0,\,0)}, g)$.

2. Charges

The space-time vector

$$\boldsymbol{c} = \begin{bmatrix} 1 \\ 1 \\ 1 \\ 1 \end{bmatrix} \tag{2.1}$$

is an eigenvector corresponding to eigenvalue $c = 1$ for each element of the group S_3 in (1.8). The eigenspace spanned by $\boldsymbol{c}$ is the principal space-time diagonal $\boldsymbol{D} = \{(D, D, D, D) \mid D\in\mathbf{R}\}$. The space-projection of $\boldsymbol{D}$ is given by $\boldsymbol{S} = \{(D, D, D, 0) \mid D\in\mathbf{R}\}$, which is the principal space diagonal. The time-projection of $\boldsymbol{D}$ is given by $\boldsymbol{T} = \{(0, 0, 0, D) \mid D\in\mathbf{R}\}$, which is the time axis. Since $(D, D, D, 0) \cdot (0, 0, 0, D) = 0$ for all D in $\boldsymbol{R}$, the principal space diagonal $\boldsymbol{S}$ and the time-axis $\boldsymbol{T}$ are mutually orthogonal in space-time.

We first consider the principal space diagonal $\boldsymbol{S}$. Let $\boldsymbol{P} = \{(X, Y, Z, 0)\in\mathbf{R}^{(3,\,1)} \mid (X, Y, Z, 0) \cdot (D, D, D, 0) = 0$ for all $(D, D, D, 0)\in\boldsymbol{S}\,\}$ denote the space plane that is perpendicular to $\boldsymbol{S}$. Then the unit space vectors (1, 0, 0, 0), (0, 1, 0, 0), (0, 0, 1, 0) project onto the space plane $\boldsymbol{P}$ as the vertices $\underline{X}$, $\underline{Y}$, $\underline{Z}$ of an equilateral triangle. Now we can see that the group S_3 in (1.8) acts on the unit space vectors as the dihedral group showing the six symmetries of the equilateral triangle $\underline{X}$, $\underline{Y}$, $\underline{Z}$ in the space plane $\boldsymbol{P}$.

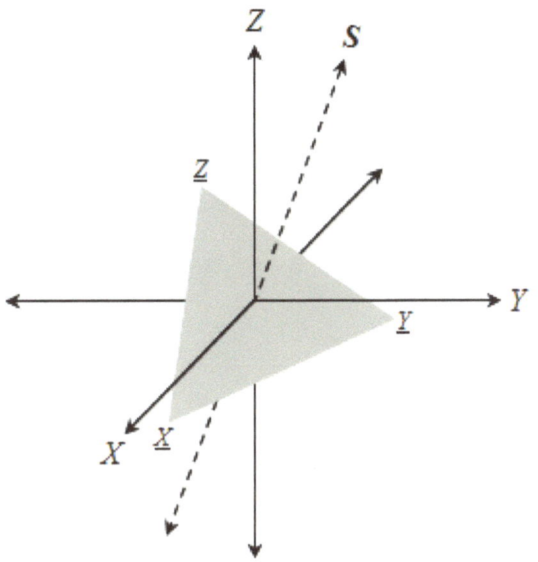

Figure 2.1. *The equilateral triangle $\underline{X}$, $\underline{Y}$, $\underline{Z}$ in the space plane **P***

We identify the space plane ***P*** with a complex plane ***C*** such that the origin (0, 0, 0, 0) in ***P*** corresponds to the origin 0 in ***C*** and the vertices $\underline{X}$, $\underline{Y}$, $\underline{Z}$ of the equilateral triangle in ***P*** correspond to the cube roots of unity 1, $e^{2\pi i/3}$, $e^{4\pi i/3}$ in ***C***, respectively.

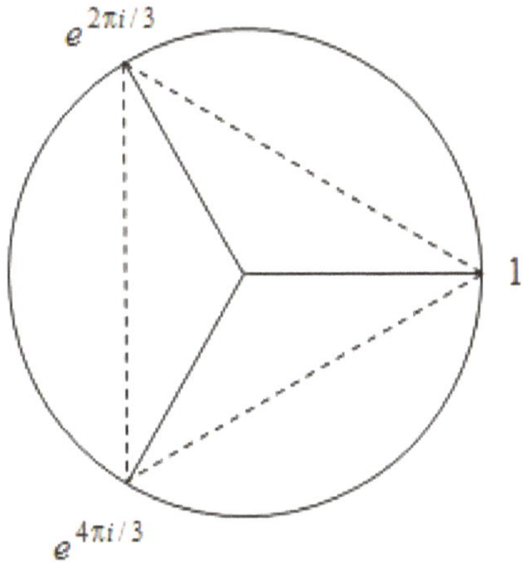

Figure 2.2. *The cube roots of unity in the complex plane **C***

Recall the definition of the map ***m***(4) on the complex plane ***C*** from **[2]**. Deform the map ***m***(4) such that the "island" of red, green and yellow regions forms an equilateral triangle with the origin 0 at its center and the cube roots of unity 1, $e^{2\pi i/3}$, $e^{4\pi i/3}$ at the centeroids of the red, green and yellow regions respectively.

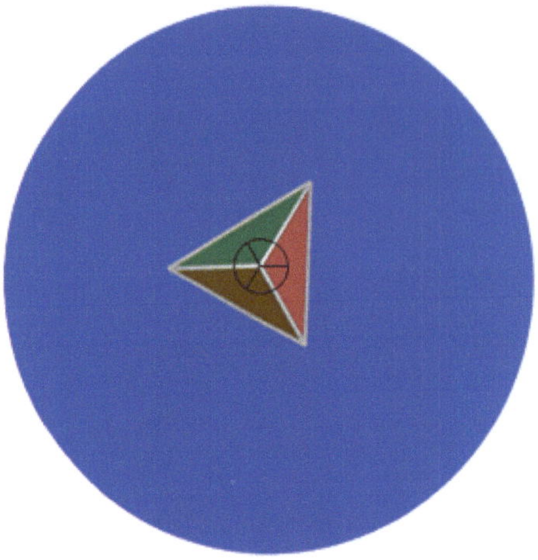

Figure 2.3. *The map* **m**(4) *with the cube roots of unity in the complex plane* ***C***

The "sea" formed by the blue region contains the point at infinity which is the "horizon" of the complex plane ***C***. Using the stereographic projection we may view the extended complex plane $\boldsymbol{C} \cup \{\infty\}$ as the Riemann sphere. Then the origin of the complex plane corresponds to the south pole of the Riemann sphere, the map **m**(4) on the complex plane is stereographically projected onto the southern hemisphere of the Riemann sphere and the point at infinity of the complex plane corresponds to the north pole of the Riemann sphere.

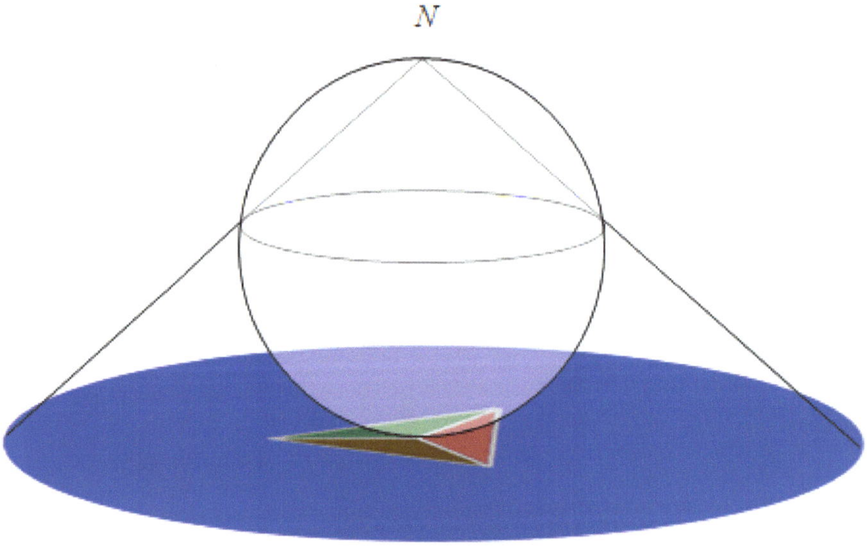

Figure 2.4. *The stereographic projection of the map* **m**(4) *on the Riemann sphere*

The electric charge space $\boldsymbol{Q} = \{Q \mid Q \in \mathbf{R}\}$ is divided into intervals of elementary charge units, where an elementary charge unit corresponds to the magnitude of the charge $\boldsymbol{e}$ on the electron. The calculation of the electromagnetic gauge coupling constant **[2]** shows that

$$
\begin{aligned}
& \hbar c/\boldsymbol{e}^2 = 137 \\
\Rightarrow\ & 1/\boldsymbol{e}^2 = 137 \quad (\text{since } \hbar = c = 1 \text{ in Planck units}) \qquad (2.2)\\
\Rightarrow\ & \boldsymbol{e} = 137^{-1/2}
\end{aligned}
$$

so that an elementary charge unit 1 in $\boldsymbol{Q}$ is $137^{-1/2}$ (or a little more than 1/12) times a Planck unit 1. The electric charge space $\boldsymbol{Q}$ under addition is isomorphic to the abelian group **R** of real numbers under addition. Consider the function $\boldsymbol{q}$ from the charge space $\boldsymbol{Q}$ into the complex$z$-plane $\boldsymbol{C}$ given by

$$\boldsymbol{q} : \boldsymbol{Q} \rightarrow \boldsymbol{C}\,;\, Q \rightarrow z = e^{2\pi i Q} \qquad (2.3)$$

The function $\boldsymbol{q}$ is an additive homomorphism because addition in the electric charge space $\boldsymbol{Q}$ corresponds to addition of angles in the complex z-plane $\boldsymbol{C}$. The image of the homomorphism $\boldsymbol{q}$ is the unit circle in the complex plane $\boldsymbol{C}$. By figure 2.2, the preimages of the cube roots of unity subdivide the the electric charge space $\boldsymbol{Q}$ into multiples of the elementary charge 1/3 as shown in figure 2.5.

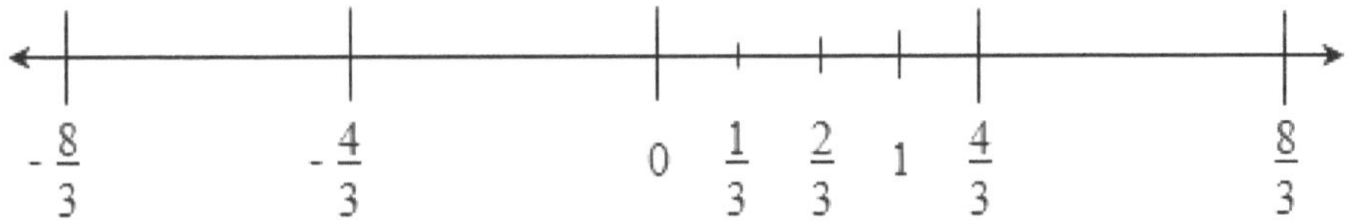

Figure 2.5. *The electric charge space* ***Q***

We now adjoin ∞ to the complex plane and extend the homomorphism

$$\boldsymbol{q}_\infty : \boldsymbol{Q} \rightarrow \boldsymbol{C} \cup \{\infty\} \qquad (2.4)$$

according to the simple closed Dirac path shown in figure 2.6.

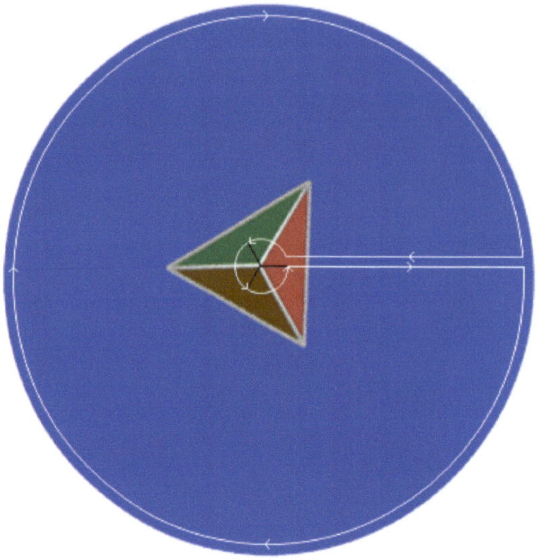

Figure 2.6. *The Dirac path on the map* $\mathbf{m}(4)$ *in the complex plane*

Explicitly, the homomorphism $\boldsymbol{q}_\infty$ is given as follows. As Q goes from 0 to 1/3, the Dirac path goes from ∞ clockwise along the horizon to the real axis and then backwards along the positive real axis to the unit circle and then anticlockwise along the unit circle to $e^{2\pi i/3}$. As Q goes from 1/3 to 2/3, the Dirac path goes anticlockwise along the unit circle from $e^{2\pi i/3}$ to $e^{4\pi i/3}$. As Q goes from 2/3 to 1, the Dirac path goes anticlockwise along the unit circle from $e^{4\pi i/3}$ to 1. Finally as Q goes from 1 to 4/3, the Dirac path goes forward from 1 along the positive real axis and then clockwise along the horizon to ∞ and its starting point on the horizon. We define the addition of two points on the Dirac path via the addition of their preimages as electric charges, i.e. $\boldsymbol{q}_\infty$ is a homomorphism. The image of the homomorphism $\boldsymbol{q}_\infty$ is the Dirac path which is isomorphic to the unitary group $\boldsymbol{U}(1)$. The kernel of the homomorphism $\boldsymbol{q}_\infty$ is the additive subgroup of the electric charge space $\boldsymbol{Q}$ isomorphic to the integers $\boldsymbol{Z} = <4/3>$ = { ..., -8/3, -4/3, 0, 4/3, 8/3, ...} generated by the elementary charge $Q = 4/3$ in the electric charge space. The homomorphism $\boldsymbol{q}_\infty$ wraps the the interval [0, 4/3) (and all its translations by multiples of 4/3) in the electric charge space $\boldsymbol{Q}$ around the circumference of the unit circle modulo 2π in the complex z-plane $\boldsymbol{C}$. The elements of the group $\boldsymbol{Q}/\boldsymbol{Z}$ are the cosets $\{ Q + \boldsymbol{Z} \mid Q \in [0, 4/3) \}$ of $\boldsymbol{Z}$ in $\boldsymbol{Q}$. Since only integer multiples of the four elementary charges 0, 1/3, 2/3, 1 are actually found in the universe, we must restrict ourselves to the subgroup $\boldsymbol{Q}_4 = \{0 + \boldsymbol{Z}, 1/3 + \boldsymbol{Z}, 2/3 + \boldsymbol{Z}, 1 + \boldsymbol{Z}\}$ while assigning electric charges to particles. This subgroup $\boldsymbol{Q}_4$ is isomorphic to the group $\boldsymbol{Z}_4 = \{\underline{0}, \underline{1}, \underline{2}, \underline{3}\}$of integers with addition modulo 4. This explains why the four colours 0, 1, 2, 3 in **[2]** are represented by the palette

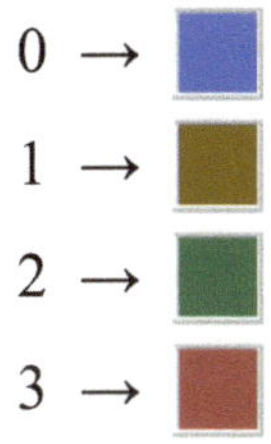

(2.5)

and the regions $\underline{0}$, $\underline{1}$, $\underline{2}$, $\underline{3}$ of the map $\boldsymbol{m}(4)$ represent the electric charges 0, 1/3, 2/3, 1 respectively. From the perspective of the Riemann sphere in figure 2.4, the north pole would correspond to Dirac's construction of the magnetic monopole. An electric charge at the south pole of the Riemann sphere which is moving in space-time must create a magnetic field perpendicular to itself, i.e. along the central axis connecting the south and north poles of the Riemann sphere. Conversely, a magnetic field due to the magnetic monopole at the north pole along the central axis of the Riemann sphere from the south to the north pole must create electric charges at the south pole. Thus, our scheme is in complete agreement with Dirac **[9][10]** and the abstract magnetic monopole indeed creates the quantized electric charges 0, 1/3, 2/3, 1. Note that magnetic monopoles cannot be observable particles in the standard model, since they are not defined as solutions of the Schrödinger wave equation on the particle frame **[2]**. However, direct comparison of our scheme with 't Hooft's model **[11]** shows that this construction explains permanent quark confinement.

The group S_3 in (1.8) acts on the electric charges 0, 1/3, 2/3, 1 as the dihedral group showing the six symmetries of the regions $\underline{0}$, $\underline{1}$, $\underline{2}$, $\underline{3}$ of the map $\boldsymbol{m}(4)$ in the complex plane $\boldsymbol{C}$. The S_3 action keeps the electric charge 0 and the blue region $\underline{0}$ of the map fixed. This corresponds to the S_3 action on the time and space unit vector projections $\underline{T}$, $\underline{X}$, $\underline{Y}$, $\underline{Z}$ keeping the time unit vector projection $\underline{T}$ fixed.

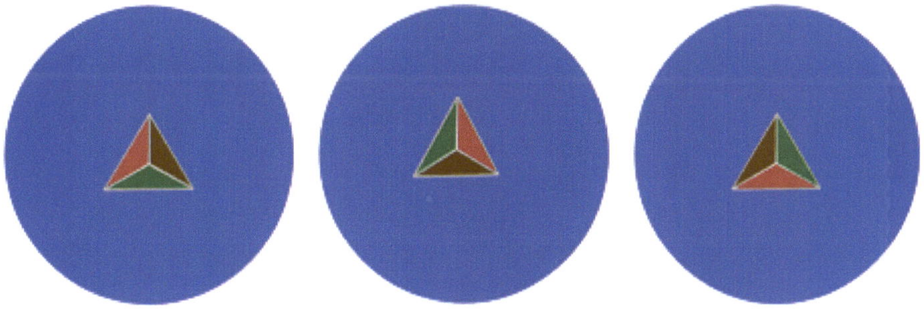

1 ρ ρ^2

(2.6)

σ $\sigma\rho$ $\sigma\rho^2$

We shall now show how the particle frame **[2]** is created at a space-time point (*X*, *Y*, *Z*, *T*) in vacuum by a spontaneous breaking of this symmetry. The Planck units along the principal space diagonal ***S*** and the time-axis ***T*** generate two copies of the integers ***Z*** under addition. These will be the scalars with respect to which we shall form certain integral group rings. There are two fundamental automorphisms of each copy of ***Z***; the identity automorphism + : ***Z*** → ***Z***; *m* → *m* and the negation automorphism - : ***Z*** → ***Z***; *m* → -*m*. The spontaneous breaking of symmetry occurs in the following sequence, corresponding to the construction of the *t*-Riemann surface in **[2]**. First, there is a translation in the space plane ***P*** that is identified with the complex plane ***C***, moving the map ***m***(4) away from the origin so that the origin lies inside the blue region $\underline{0}$. The six broken symmetries (2.6) are now displayed on six translated copies of the space diagonal ***P*** and the corresponding six copies of ***Z***. Then the two automorphisms + and - of the integers ***Z*** associated with the principal space diagonal ***S*** produce twelve broken symmetry configurations of the map, corresponding to the upper and lower halves of a sheet. Finally, the two automorphisms + and - of the integers ***Z*** associated with the time-axis ***T*** produce twenty-four broken symmetry configurations of the map corresponding to the upper and lower sheets of the *t*-Riemann surface.

Figure 2.7. *The t-Riemann surface*

The regions of the maps on the t-Riemann surface are labeled by elements of the split extension $\boldsymbol{Z}(\boldsymbol{Z}_4]S_3)]\boldsymbol{Z}S_3$ according to the scheme given in **[2]**. All the particles of the standard model are defined by selecting particular regions of the maps on the t-Riemann surface according to well-defined rules given in **[2]**. Each of the 24 Schrödinger discs containing the maps carry the value $\Psi(X, Y, Z, T)$ of the Schrödinger wave function on their boundaries which together form the outer rim of the particle frame. In particular, the two automorphisms +, - as the generators +1, -1 of the scalars $\boldsymbol{Z}$ in the group algebras provide the signs for all charges. We now see that this selection process can be explicitly defined via the Dirac path in figure 2.6 on each Schrödinger disc. The t-Riemann surface without any particles selected is called the particle frame and corresponds to the space-time point (X, Y, Z, T) in vacuum. Note that the Higgs particle corresponds to the branch point 0 of the particle frame **[2]** with respect to which the symmetry has been spontaneously broken. Thus, the spontaneous breaking of symmetry in vacuum corresponds exactly to the Higgs-Kibble mechanism **[12]** which assigns rest masses to all the particles of the standard model.

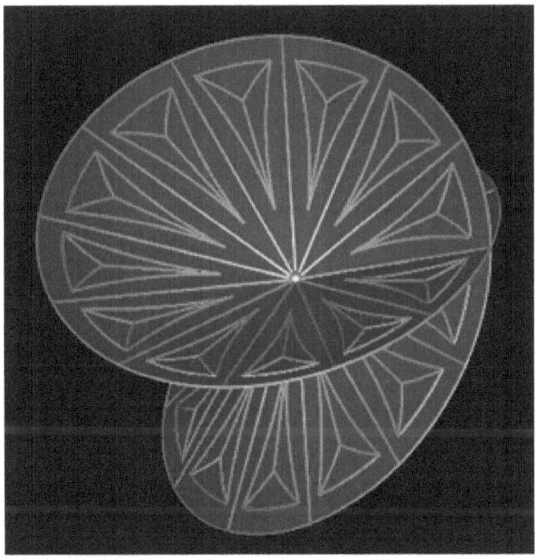

Figure 2.8. *The particle frame*

The embedding of the particle frame in space-time at (X, Y, Z, T) takes a Planck unit of time. As ΔT goes from 0 to 1 in a unit interval of Planck time, the parameter t goes goes from 0 to 4π embedding the full t-Riemann surface and particle frame in space-time at $(X, Y, Z, T + 1)$. There are 24 Dirac paths that can select any set of the regions on the particle frame and all the 24 Dirac paths together can be parameterized by t. Thus, the electric charge space $\boldsymbol{Q}$ and the group $\boldsymbol{Z}_4 = \{\underline{0}, \underline{1}, \underline{2}, \underline{3}\}$ can be realized purely in terms of time translations within a unit interval of Planck time.

3. Chiralities

The time-axis $\boldsymbol{T} = \{T = (0, 0, 0, T) \mid T \in \mathbf{R}\}$ is divided into intervals of Planck time units. The cosmological time-line **[2]** is superposed along the positive time-axis and the time interval [0, 1) corresponds to the Planck epoch. Consider the function $\boldsymbol{p}$ from the time-axis $\boldsymbol{T}$ into the complex z-plane $\boldsymbol{C}$ given by

$$\boldsymbol{p} : \boldsymbol{T} \rightarrow \boldsymbol{C} ; T \rightarrow z = e^{2\pi i T} \tag{3.1}$$

The function $\boldsymbol{p}$ is an additive homomorphism because addition on the time-axis $\boldsymbol{T}$ corresponds to addition of angles in the complex z-plane $\boldsymbol{C}$. The image of the homomorphism $\boldsymbol{p}$ is isomorphic to the unitary group $\boldsymbol{U}(1)$. The kernel of the homomorphism $\boldsymbol{p}$ is the additive subgroup of $\boldsymbol{T}$ isomorphic to the integers $\boldsymbol{Z} = \langle 1 \rangle = \{ ..., -2, -1, 0, 1, 2, ...\}$ generated by the unit Planck time $T = 1$ on the time-axis. The homomorphism $\boldsymbol{p}$ wraps the Planck intervals on the time-axis $\boldsymbol{T}$ around the circumference of the unit circle modulo 2π in the complex z-plane $\boldsymbol{C}$. The quotient group $\boldsymbol{T/Z}$ is isomorphic to the image $\boldsymbol{U}(1)$. The elements of $\boldsymbol{T/Z}$ are the cosets $\{ T + \boldsymbol{Z} \mid T \in [0,1) \}$ of $\boldsymbol{Z}$ in $\boldsymbol{T}$ showing how the unit Planck interval is replicated along the entire time-axis.

During the Planck epoch, the particle frame is embedded in space-time as the t-Riemann surface **[2]** Section 8.1.4 via the function

$$\boldsymbol{C} \rightarrow \boldsymbol{C} ; z \rightarrow t = z^2 \tag{3.2}$$

There are four Schrödinger discs on the particle frame during the Planck epoch, carrying the four wave functions Ψ_0, Ψ_1, Ψ_2, Ψ_3 respectively:

Wave Function	t-Riemann surface	z-plane	Time-axis $\boldsymbol{T}$
Ψ_0	$0 \leq arg\, t < \pi$	$0 \leq arg\, z < \pi/2$	[0, 1/4)
Ψ_1	$\pi \leq arg\, t < 2\pi$	$\pi/2 \leq arg\, z < \pi$	[1/4, 1/2)
Ψ_2	$2\pi \leq arg\, t < 3\pi$	$\pi \leq arg\, z < 3\pi/2$	[1/2, 3/4)
Ψ_3	$3\pi \leq arg\, t < 4\pi$	$3\pi/2 \leq arg\, z < \pi$	[3/4, 1)

(3.3)

Thus, the four Planck epoch wave functions $\{\Psi_0, \Psi_1, \Psi_2, \Psi_3\}$ yield the fourth roots of unity subgroup $\{1, i, -1, -i\}$ of $\boldsymbol{U}(1)$ in the z-plane and the addition modulo 4 subgroup $\mathbf{Z}_4 = \{\underline{0}, \underline{1}, \underline{2}, \underline{3}\}$ of $\boldsymbol{T/Z}$, where

$$\begin{aligned} \underline{0} &= 0 + \boldsymbol{Z} \\ \underline{1} &= 1/4 + \boldsymbol{Z} \\ \underline{2} &= 1/2 + \boldsymbol{Z} \\ \underline{3} &= 3/4 + \boldsymbol{Z} \end{aligned} \tag{3.4}$$

by (3.1), (3.2) and (3.3).

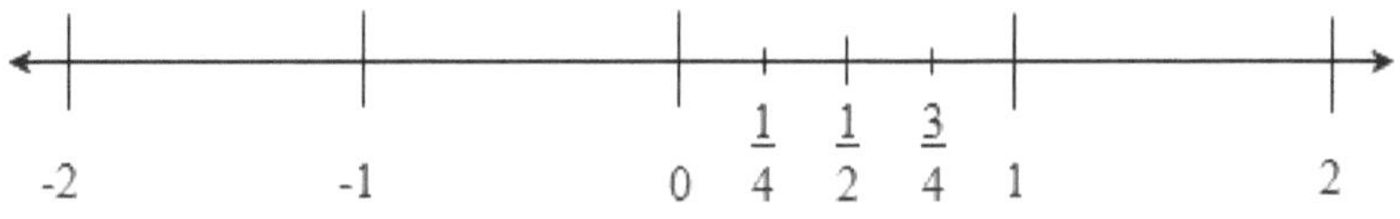

Figure 3.1. *The time-axis* ***T***

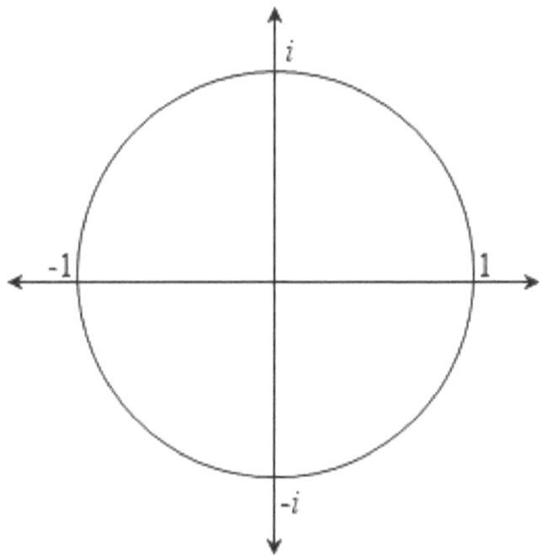

Figure 3.2. *The z-plane* ***C***

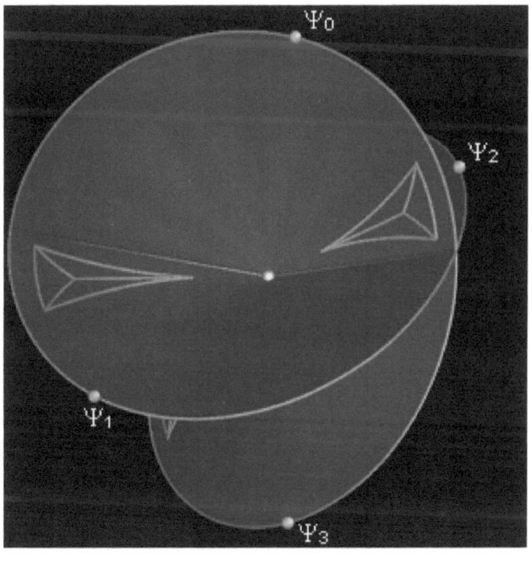

Figure 3.3. *The four wave functions on the particle frame during the Planck epoch*

This is the true motivation for considering a vector wave function $\mathbf{\Psi}$ with four components

$$\mathbf{\Psi} = \begin{bmatrix} \Psi_0 \\ \Psi_1 \\ \Psi_2 \\ \Psi_3 \end{bmatrix} \tag{3.5}$$

which is a solution of Dirac's relativistic wave equation **[13]**

$$(i\hbar \frac{\partial}{\partial T} + i\hbar (\gamma_X \frac{\partial}{\partial X} + \gamma_Y \frac{\partial}{\partial Y} + \gamma_Z \frac{\partial}{\partial Z}) - \gamma_T mc^2) \mathbf{\Psi} = 0 \tag{3.6}$$

with the requirement that any solution of (3.6) is also a solution of the relativistic Schrödinger wave equation (the converse need not be true). With this requirement, we can solve (3.6) explicitly for $\gamma_X, \gamma_Y, \gamma_Z, \gamma_T$:

$$\gamma_X = \begin{bmatrix} 0 & 0 & 0 & 1 \\ 0 & 0 & 1 & 0 \\ 0 & -1 & 0 & 0 \\ -1 & 0 & 0 & 0 \end{bmatrix} \quad \gamma_Y = \begin{bmatrix} 0 & 0 & 0 & -i \\ 0 & 0 & i & 0 \\ 0 & i & 0 & 0 \\ -i & 0 & 0 & 0 \end{bmatrix} \quad \gamma_Z = \begin{bmatrix} 0 & 0 & 1 & 0 \\ 0 & 0 & 0 & -1 \\ -1 & 0 & 0 & 0 \\ 0 & 1 & 0 & 0 \end{bmatrix}$$
$$\gamma_T = \begin{bmatrix} 0 & 0 & 1 & 0 \\ 0 & 0 & 0 & 1 \\ 1 & 0 & 0 & 0 \\ 0 & 1 & 0 & 0 \end{bmatrix} \tag{3.7}$$

In Einstein's tensor notation, the four gamma matrices are traditionally written as $\gamma_T = \gamma^0$, $\gamma_X = \gamma^1$, $\gamma_Y = \gamma^2$, $\gamma_Z = \gamma^3$. The fifth gamma matrix is defined as

$$\gamma^5 = i\,\gamma^0\gamma^1\gamma^2\gamma^3 = \begin{bmatrix} -1 & 0 & 0 & 0 \\ 0 & -1 & 0 & 0 \\ 0 & 0 & 1 & 0 \\ 0 & 0 & 0 & 1 \end{bmatrix} \tag{3.8}$$

Then the left and right chirality of the wave function $\mathbf{\Psi}$ is defined in terms of the two matrices

$$\frac{1 - \gamma^5}{2} = \frac{1}{2}\begin{bmatrix} 2 & 0 & 0 & 0 \\ 0 & 2 & 0 & 0 \\ 0 & 0 & 0 & 0 \\ 0 & 0 & 0 & 0 \end{bmatrix} = \begin{bmatrix} 1 & 0 & 0 & 0 \\ 0 & 1 & 0 & 0 \\ 0 & 0 & 0 & 0 \\ 0 & 0 & 0 & 0 \end{bmatrix} \tag{3.9}$$

$$\frac{1+\gamma^5}{2} = \frac{1}{2}\begin{bmatrix} 0 & 0 & 0 & 0 \\ 0 & 0 & 0 & 0 \\ 0 & 0 & 2 & 0 \\ 0 & 0 & 0 & 2 \end{bmatrix} = \begin{bmatrix} 0 & 0 & 0 & 0 \\ 0 & 0 & 0 & 0 \\ 0 & 0 & 1 & 0 \\ 0 & 0 & 0 & 1 \end{bmatrix}$$

as follows:

$$\mathbf{\Psi_L} = \frac{1-\gamma^5}{2}\mathbf{\Psi} = \begin{bmatrix} 1 & 0 & 0 & 0 \\ 0 & 1 & 0 & 0 \\ 0 & 0 & 0 & 0 \\ 0 & 0 & 0 & 0 \end{bmatrix}\begin{bmatrix} \Psi_0 \\ \Psi_1 \\ \Psi_2 \\ \Psi_3 \end{bmatrix} = \begin{bmatrix} \Psi_0 \\ \Psi_1 \\ 0 \\ 0 \end{bmatrix}$$

$$\mathbf{\Psi_R} = \frac{1+\gamma^5}{2}\mathbf{\Psi} = \begin{bmatrix} 0 & 0 & 0 & 0 \\ 0 & 0 & 0 & 0 \\ 0 & 0 & 1 & 0 \\ 0 & 0 & 0 & 1 \end{bmatrix}\begin{bmatrix} \Psi_0 \\ \Psi_1 \\ \Psi_2 \\ \Psi_3 \end{bmatrix} = \begin{bmatrix} 0 \\ 0 \\ \Psi_2 \\ \Psi_3 \end{bmatrix} \tag{3.10}$$

By the standard model completion rule **[2]**, superposition of the particle frames of all types of fermions during the present epoch shows that they fit perfectly on the particle frame according to their labels. Each of the 24 Schrödinger discs of the particle frame holds a unique type of fermion.

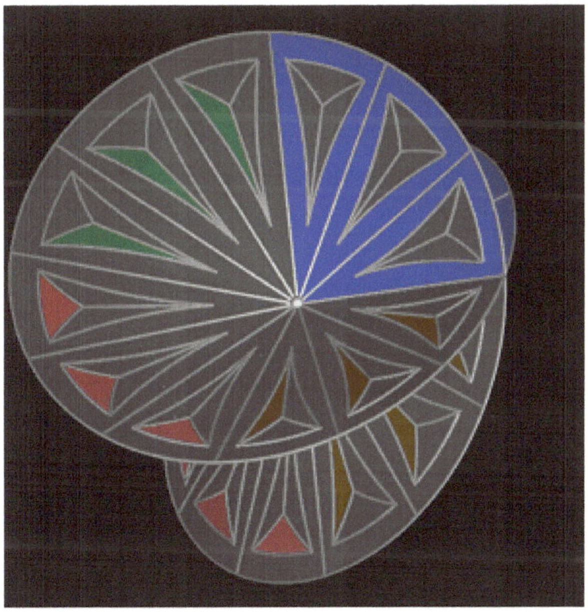

Figure 3.4. *The perfect fitting of the fermions in the standard model*

The positions of the 24 fermions on the upper and lower sheets of the particle frame are shown below.

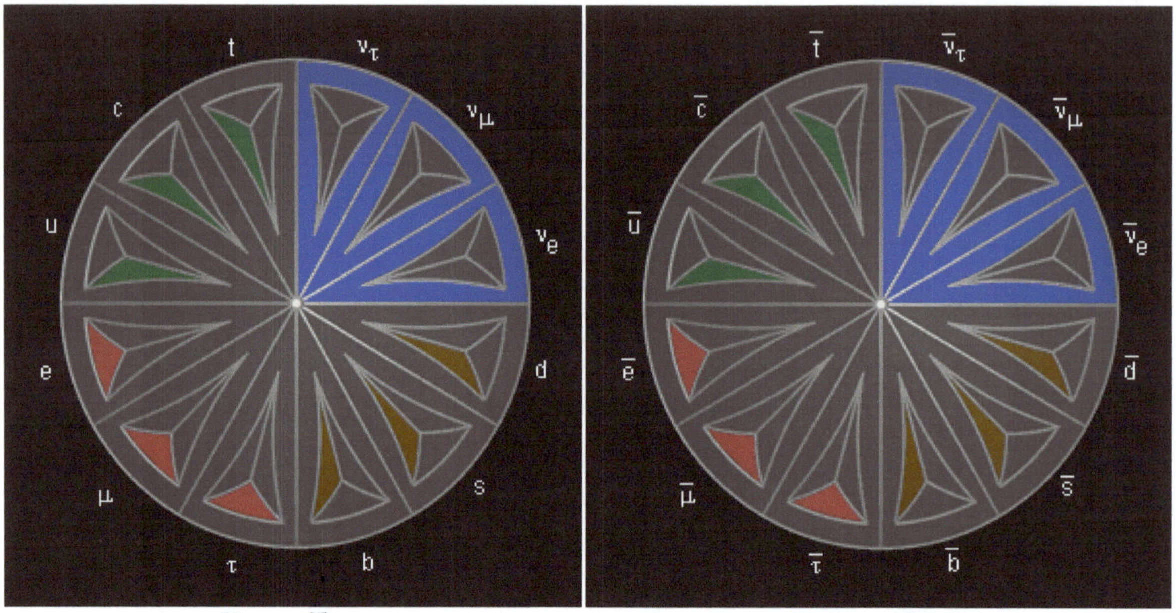

Upper Sheet
Left-handed fermions

Lower Sheet
Right-handed fermions

Figure 3.5. *The 24 fermions on the particle frame*

In particular, figure 3.5 now completely justifies the formulation of the weak isospin rule **[3]** in terms of the left-handed and right-handed chiralities of the fermion particles and antiparticles, respectively. Note that the four components Ψ_0, Ψ_1, Ψ_2, Ψ_3 of the wave function $\boldsymbol{\Psi}$ correspond to the addition modulo 4 subgroup $\mathbf{Z}_4 = \{\underline{0}, \underline{1}, \underline{2}, \underline{3}\}$ of $\boldsymbol{T/Z}$ by (3.4). Thus, each component is obtained by *currying* **[14]** the wave function in a Planck time interval:

$$\begin{aligned} \boldsymbol{\Psi}(X, Y, Z, \underline{0}) &= \Psi_0(X, Y, Z) \\ \boldsymbol{\Psi}(X, Y, Z, \underline{1}) &= \Psi_1(X, Y, Z) \\ \boldsymbol{\Psi}(X, Y, Z, \underline{2}) &= \Psi_2(X, Y, Z) \\ \boldsymbol{\Psi}(X, Y, Z, \underline{3}) &= \Psi_3(X, Y, Z) \end{aligned} \tag{3.11}$$

The S_3 parts of the $\boldsymbol{Z}_4]S_3$ labels of the fermions correspond to the $S_3 = \{1, \rho, \rho^2, \sigma, \sigma\rho, \sigma\rho^2\}$ subgroup of the Lorentz group $\mathbf{O}^{(3,\,1)}$ by (1.8) and (1.9). Since Dirac's relativistic wave equation (3.6) is invariant under Lorentz transformations, we can replicate each component of the wave function by the 6 elements of S_3 without affecting the fact that it is a solution of (3.6). Then, the wave function $\boldsymbol{\Psi}$ specifies the individual wave functions of each of the 24 fermions according to the following table:

$\Psi_0(Y, X, Z)$	$\Psi_0(Z, Y, X)$	$\Psi_0(X, Z, Y)$	$\Psi_0(Z, X, Y)$	$\Psi_0(Y, Z, X)$	$\Psi_0(X, Y, Z)$
$\Psi_1(Y, X, Z)$	$\Psi_1(Z, Y, X)$	$\Psi_1(X, Z, Y)$	$\Psi_1(Z, X, Y)$	$\Psi_1(Y, Z, X)$	$\Psi_1(X, Y, Z)$
$\Psi_2(Y, X, Z)$	$\Psi_2(Z, Y, X)$	$\Psi_2(X, Z, Y)$	$\Psi_2(Z, X, Y)$	$\Psi_2(Y, Z, X)$	$\Psi_2(X, Y, Z)$
$\Psi_3(Y, X, Z)$	$\Psi_3(Z, Y, X)$	$\Psi_3(X, Z, Y)$	$\Psi_3(Z, X, Y)$	$\Psi_3(Y, Z, X)$	$\Psi_3(X, Y, Z)$

Figure 3.6. *The curried wave function table of fermions*

For each fermion F, we define its *time-chirality* $\chi_T(F)$ and its *space-chirality* $\chi_S(F)$ as follows:

$$\chi_T(F) = \begin{cases} +1, & \text{if the curried component } \Psi_0 \text{ or } \Psi_1 \text{ of its wave function } \boldsymbol{\Psi} \text{ is obtained in the left-half of the Planck time interval, i.e. } \boldsymbol{\Psi}_L \text{ by (3.10)} \\ -1, & \text{if the curried component } \Psi_2 \text{ or } \Psi_3 \text{ of its wave function } \boldsymbol{\Psi} \text{ is obtained in the right-half of the Planck time interval, i.e. } \boldsymbol{\Psi}_R \text{ by (3.10)} \end{cases}$$

$$\chi_S(F) = \begin{cases} -1, & \text{if the argument } X, Y, Z \text{ of its wave function } \boldsymbol{\Psi} \text{ is obtained in a left-handed space coordinate system } \sigma, \sigma\rho, \sigma\rho^2 \text{ by (1.9)} \\ +1, & \text{if the argument } X, Y, Z \text{ of its wave function } \boldsymbol{\Psi} \text{ is obtained in a right-handed space coordinate system } 1, \rho, \rho^2 \text{ by (1.9)} \end{cases} \qquad (3.12)$$

Thus, we obtain the *space-time table* of the 24 fermions:

Planck Time Intervals		*Left Space-Chirality*			*Right Space-Chirality*			
0	*Left*	u	c	t	ν_τ	ν_μ	ν_e	
1	*Time-Chirality*	d	s	b	τ	μ	e	*Particles*
2	*Right*	$\bar{u}$	$\bar{c}$	$\bar{t}$	$\bar{\nu}_\tau$	$\bar{\nu}_\mu$	$\bar{\nu}_e$	*Antiparticles*
3	*Time-Chirality*	$\bar{d}$	$\bar{s}$	$\bar{b}$	$\bar{\tau}$	$\bar{\mu}$	$\bar{e}$	
			Quarks			*Leptons*		

Figure 3.7. *The space-time table of fermions*

We can now write the dual *space-charge table* of the 24 fermions:

Unsigned Electric Charge		*Left Time-Chirality*			*Right Time-Chirality*			
2	*Left*	u	c	t	$\bar{u}$	$\bar{c}$	$\bar{t}$	
1	*Space-Chirality*	d	s	b	$\bar{d}$	$\bar{s}$	$\bar{b}$	*Quarks*
0	*Right*	ν_τ	ν_μ	ν_e	$\bar{\nu}_\tau$	$\bar{\nu}_\mu$	$\bar{\nu}_e$	*Leptons*
3	*Space-Chirality*	τ	μ	e	$\bar{\tau}$	$\bar{\mu}$	$\bar{e}$	
			Particles			*Antiparticles*		

Figure 3.8. *The space-charge table of fermions*

The tables show the duality between time and charge and the relation between space-chirality which causes the quark/lepton divide and time-chirality which causes the matter/antimatter divide.

Dirac's wave equation (3.6) has a mass term m. Working in Planck units we have $\hbar = c = 1$ and therefore m is also given in Planck units of mass:

$$(i\frac{\partial}{\partial T} + i\,(\gamma_X\frac{\partial}{\partial X} + \gamma_Y\frac{\partial}{\partial Y} + \gamma_Z\frac{\partial}{\partial Z}) - \gamma_T\, m)\, \boldsymbol{\Psi} = 0 \qquad (3.13)$$

By the curried wave function of figure 3.6, we can now see how $\boldsymbol{\Psi}$ explicitly assigns the masses of all 24 fermions. The scalar Higgs field φ interacts with the curried wave function $\boldsymbol{\Psi}$ of each fermion via the *Yukawa coupling* **[15]**

$$V = g\, \boldsymbol{\Psi}\, \varphi\, \boldsymbol{\Psi} \qquad (3.14)$$

The mass term m of $\boldsymbol{\Psi}$ for each fermion can now be associated in a one-to-one correspondence with the permutation $\psi \in sym(\boldsymbol{Z}_4]S_3)$ that defines the rest mass of the fermion according to the mass rule in **[2][3]**. The labels for the fermions occur in pairs of the form

$$\left(\pm(\underline{m}, \alpha)\downarrow\begin{pmatrix}\psi\\ \psi^{\mu}\end{pmatrix}, \beta+\gamma\right) \quad \text{and} \quad \left(\pm(\underline{m}, \alpha)\uparrow\begin{pmatrix}\psi\\ \psi^{\mu}\end{pmatrix}, \beta+\gamma\right) \qquad (3.15)$$

on the particle frame **[1][2]**. The $\uparrow$ and $\downarrow$ group actions in (3.15) specify a pair of superposed left-handed and right-handed Schrödinger discs on the particle frame, with respect to time-chirality of fermions. Thus, the Yukawa coupling (3.14) can be interpreted as the binding potential between a left-handed and right-handed fermion on the particle frame which attributes mass m to the fermion.

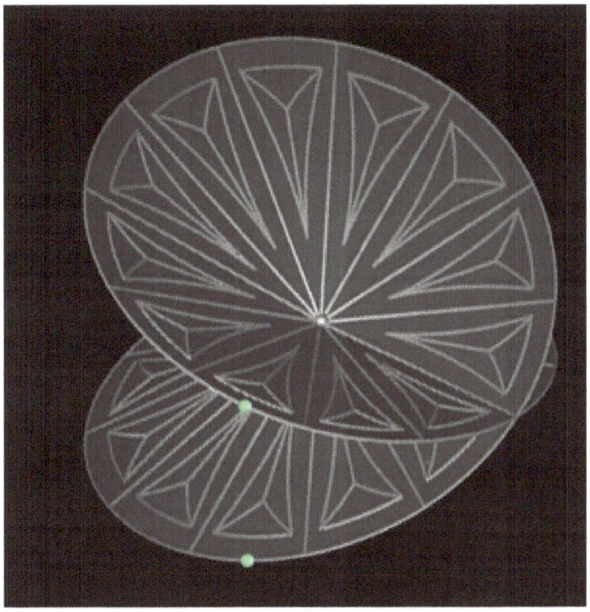

Figure 3.9. *The Yukawa coupling of a left-handed and right-handed fermion*

Note that the mass m of the fermion obtained by the Yukawa coupling is not constant and varies with the energy scale at which it is measured. The rest mass m of the fermion is obtained after renormalization and interaction of the Yukawa coupling with the Higgs field according to (3.14). This is the wave function form of the Higgs-Kibble mechanism due to the spontaneous breaking of symmetry and creation of the particle frame in vacuum **[16]**.

4. Representations

We shall now show how all the particles of the standard model correspond to irreducible representations of the Poincaré group $\mathbf{I}^{(3,\ 1)}$, according to Wigner's classification **[17]**. The Poincaré group $\mathbf{I}^{(3,\ 1)}$ has ten infinitesimal Lie group generators that are traditionally written in Einstein's tensor notation as:

$$\begin{array}{lll} P_T = P^0 & & \\ P_X = P^1 & N_X = M^{01} & J_X = M^{32} \\ P_Y = P^2 & N_Y = M^{02} & J_Y = M^{13} \\ P_Z = P^3 & N_Z = M^{03} & J_Z = M^{21} \end{array} \tag{4.1}$$

The generators $\boldsymbol{P} = (P_T, P_X, P_Y, P_Z)$ correspond to infinitesimal time and space translations respectively, the generators $\boldsymbol{N} = (N_X, N_Y, N_Z)$ correspond to infinitesimal

Lorentz boosts along the space axes and the generators $\boldsymbol{J} = (J_X, J_Y, J_Z)$ correspond to the three infinitesimal components of angular momentum along the space axes. The *Pauli vector* $\boldsymbol{w}$ is defined by **[18]**:

$$\begin{aligned} w_T &= P_X J_X + P_Y J_Y + P_Z J_Z \\ w_X &= P_Y N_Z - P_Z N_Y - P_T J_X \\ w_Y &= P_Z N_X - P_X N_Z - P_T J_Y \\ w_Z &= P_X N_Y - P_Y N_X - P_T J_Z \end{aligned} \tag{4.2}$$

Now, the Poincaré group $\mathbf{I}^{(3,\,1)}$ has two *Casimir operators* **[19]**:

$$\begin{aligned} \boldsymbol{P}^2 &= P_X^{\,2} + P_Y^{\,2} + P_Z^{\,2} - P_T^{\,2} \\ \boldsymbol{w}^2 &= w_X^{\,2} + w_Y^{\,2} + w_Z^{\,2} - w_T^{\,2} \end{aligned} \tag{4.3}$$

The irreducible unitary representations of the Poincaré group $\mathbf{I}^{(3,\,1)}$ are classified according to the eigenvalues of the Casimir operators $\boldsymbol{P}^2$ and $\boldsymbol{w}^2$. In many cases, the irreducible unitary representations do not correspond to physical particles. There are only four cases in which the irreducible unitary representations actually correspond to physical particles:

$$\begin{aligned} &(1) \quad \boldsymbol{P}^2 = m^2 > 0 \quad \text{and} \quad P_T > 0 \\ &(2) \quad \boldsymbol{P}^2 = m^2 > 0 \quad \text{and} \quad P_T < 0 \\ &(3) \quad \boldsymbol{P}^2 = m^2 = 0 \quad \text{and} \quad P_T > 0 \\ &(4) \quad \boldsymbol{P}^2 = m^2 = 0 \quad \text{and} \quad P_T < 0 \end{aligned} \tag{4.4}$$

There is always a Lorentz transformation of space-time with respect to which a particle is at rest. In this rest frame of reference, the eigenvalues that correspond to the irreducible unitary representation of the particle are m^2 and $s(s+1)$, where m is the rest mass and s is the spin of the particle. Let $\mathbf{CI}^{(3,\,1)}$ denote the complex group algebra of the Poincaré group $\mathbf{I}^{(3,\,1)}$. A module M over $\mathbf{CI}^{(3,\,1)}$ is called *simple* if its only submodules are $\{0\}$ and M. Following Noether **[20]**, each irreducible unitary representation of the Poincaré group $\mathbf{I}^{(3,\,1)}$ is equivalent to a simple module M over the complex group algebra of the Poincaré group $\mathbf{CI}^{(3,\,1)}$.

We shall now build the irreducible unitary representations corresponding to each particle in the standard model as defined in **[2]**. Note that all particles of the standard model are defined by Schrödinger discs which are labeled by elements of the split extension $\boldsymbol{Z}(\boldsymbol{Z}_4]S_3)]\boldsymbol{Z}S_3$ of the form

$$(\pm(\underline{m}, \alpha), \beta+\gamma) \tag{4.5}$$

where $(\underline{m}, \alpha) \in \boldsymbol{Z}_4]S_3$, $\beta \in S_3$ and $\gamma \in S_3$. First, write a faithful representation of S_3 corresponding to its permutation action on the basis {(1, 0, 0), (0, 1, 0), (0, 0, 1)} of space $\mathbf{R}^3$ given by the top-left 3 × 3 blocks of the six matrices in (1.8), as follows:

$$1 = \begin{bmatrix} 1 & 0 & 0 \\ 0 & 1 & 0 \\ 0 & 0 & 1 \end{bmatrix} \quad \rho = \begin{bmatrix} 0 & 1 & 0 \\ 0 & 0 & 1 \\ 1 & 0 & 0 \end{bmatrix} \quad \rho^2 = \begin{bmatrix} 0 & 0 & 1 \\ 1 & 0 & 0 \\ 0 & 1 & 0 \end{bmatrix}$$

$$\sigma = \begin{bmatrix} 0 & 1 & 0 \\ 1 & 0 & 0 \\ 0 & 0 & 1 \end{bmatrix} \quad \sigma\rho = \begin{bmatrix} 0 & 0 & 1 \\ 0 & 1 & 0 \\ 1 & 0 & 0 \end{bmatrix} \quad \sigma\rho^2 = \begin{bmatrix} 1 & 0 & 0 \\ 0 & 0 & 1 \\ 0 & 1 & 0 \end{bmatrix} \tag{4.6}$$

Next, write a faithful representation of the group $\boldsymbol{Z}_4$ by the four 1 × 1 matrices

$$[1], [i], [-1], [-i] \tag{4.7}$$

Putting together (4.6) and (4.7) by blocks, we obtain a faithful representation of the split extension $\boldsymbol{Z}_4]S_3$ as twenty-four 4 × 4 matrices:

$$(\underline{0}, 1) = \begin{bmatrix} 1 & 0 & 0 & 0 \\ 0 & 1 & 0 & 0 \\ 0 & 0 & 1 & 0 \\ 0 & 0 & 0 & 1 \end{bmatrix} \quad (\underline{0}, \rho) = \begin{bmatrix} 0 & 1 & 0 & 0 \\ 0 & 0 & 1 & 0 \\ 1 & 0 & 0 & 0 \\ 0 & 0 & 0 & 1 \end{bmatrix} \quad (\underline{0}, \rho^2) = \begin{bmatrix} 0 & 0 & 1 & 0 \\ 1 & 0 & 0 & 0 \\ 0 & 1 & 0 & 0 \\ 0 & 0 & 0 & 1 \end{bmatrix}$$

$$(\underline{0}, \sigma) = \begin{bmatrix} 0 & 1 & 0 & 0 \\ 1 & 0 & 0 & 0 \\ 0 & 0 & 1 & 0 \\ 0 & 0 & 0 & 1 \end{bmatrix} \quad (\underline{0}, \sigma\rho) = \begin{bmatrix} 0 & 0 & 1 & 0 \\ 0 & 1 & 0 & 0 \\ 1 & 0 & 0 & 0 \\ 0 & 0 & 0 & 1 \end{bmatrix} \quad (\underline{0}, \sigma\rho^2) = \begin{bmatrix} 1 & 0 & 0 & 0 \\ 0 & 0 & 1 & 0 \\ 0 & 1 & 0 & 0 \\ 0 & 0 & 0 & 1 \end{bmatrix} \tag{4.8}$$

$$(\underline{1}, 1) = \begin{bmatrix} 1 & 0 & 0 & 0 \\ 0 & 1 & 0 & 0 \\ 0 & 0 & 1 & 0 \\ 0 & 0 & 0 & i \end{bmatrix} \quad (\underline{1}, \rho) = \begin{bmatrix} 0 & 1 & 0 & 0 \\ 0 & 0 & 1 & 0 \\ 1 & 0 & 0 & 0 \\ 0 & 0 & 0 & i \end{bmatrix} \quad (\underline{1}, \rho^2) = \begin{bmatrix} 0 & 0 & 1 & 0 \\ 1 & 0 & 0 & 0 \\ 0 & 1 & 0 & 0 \\ 0 & 0 & 0 & i \end{bmatrix}$$

$$(\underline{1}, \sigma) = \begin{bmatrix} 0 & 1 & 0 & 0 \\ 1 & 0 & 0 & 0 \\ 0 & 0 & 1 & 0 \\ 0 & 0 & 0 & i \end{bmatrix} \quad (\underline{1}, \sigma\rho) = \begin{bmatrix} 0 & 0 & 1 & 0 \\ 0 & 1 & 0 & 0 \\ 1 & 0 & 0 & 0 \\ 0 & 0 & 0 & i \end{bmatrix} \quad (\underline{1}, \sigma\rho^2) = \begin{bmatrix} 1 & 0 & 0 & 0 \\ 0 & 0 & 1 & 0 \\ 0 & 1 & 0 & 0 \\ 0 & 0 & 0 & i \end{bmatrix}$$

$$(\underline{2}, 1) = \begin{bmatrix} 1 & 0 & 0 & 0 \\ 0 & 1 & 0 & 0 \\ 0 & 0 & 1 & 0 \\ 0 & 0 & 0 & -1 \end{bmatrix} \quad (\underline{2}, \rho) = \begin{bmatrix} 0 & 1 & 0 & 0 \\ 0 & 0 & 1 & 0 \\ 1 & 0 & 0 & 0 \\ 0 & 0 & 0 & -1 \end{bmatrix} \quad (\underline{2}, \rho^2) = \begin{bmatrix} 0 & 0 & 1 & 0 \\ 1 & 0 & 0 & 0 \\ 0 & 1 & 0 & 0 \\ 0 & 0 & 0 & -1 \end{bmatrix}$$

$$(\underline{2}, \sigma) = \begin{bmatrix} 0 & 1 & 0 & 0 \\ 1 & 0 & 0 & 0 \\ 0 & 0 & 1 & 0 \\ 0 & 0 & 0 & -1 \end{bmatrix} \quad (\underline{2}, \sigma\rho) = \begin{bmatrix} 0 & 0 & 1 & 0 \\ 0 & 1 & 0 & 0 \\ 1 & 0 & 0 & 0 \\ 0 & 0 & 0 & -1 \end{bmatrix} \quad (\underline{2}, \sigma\rho^2) = \begin{bmatrix} 1 & 0 & 0 & 0 \\ 0 & 0 & 1 & 0 \\ 0 & 1 & 0 & 0 \\ 0 & 0 & 0 & -1 \end{bmatrix}$$

$$(\underline{3}, 1) = \begin{bmatrix} 1 & 0 & 0 & 0 \\ 0 & 1 & 0 & 0 \\ 0 & 0 & 1 & 0 \\ 0 & 0 & 0 & -i \end{bmatrix} \quad (\underline{3}, \rho) = \begin{bmatrix} 0 & 1 & 0 & 0 \\ 0 & 0 & 1 & 0 \\ 1 & 0 & 0 & 0 \\ 0 & 0 & 0 & -i \end{bmatrix} \quad (\underline{3}, \rho^2) = \begin{bmatrix} 0 & 0 & 1 & 0 \\ 1 & 0 & 0 & 0 \\ 0 & 1 & 0 & 0 \\ 0 & 0 & 0 & -i \end{bmatrix}$$

$$(\underline{3}, \sigma) = \begin{bmatrix} 0 & 1 & 0 & 0 \\ 1 & 0 & 0 & 0 \\ 0 & 0 & 1 & 0 \\ 0 & 0 & 0 & -i \end{bmatrix} \quad (\underline{3}, \sigma\rho) = \begin{bmatrix} 0 & 0 & 1 & 0 \\ 0 & 1 & 0 & 0 \\ 1 & 0 & 0 & 0 \\ 0 & 0 & 0 & -i \end{bmatrix} \quad (\underline{3}, \sigma\rho^2) = \begin{bmatrix} 1 & 0 & 0 & 0 \\ 0 & 0 & 1 & 0 \\ 0 & 1 & 0 & 0 \\ 0 & 0 & 0 & -i \end{bmatrix}$$

For the fermions **[2]**, we can now write twenty-four 10×10 matrices blockwise using (4.5), (4.6), (4.7) and (4.8):

$$\nu_e = \begin{bmatrix} 1&0&0&0&0&0&0&0&0&0\\ 0&1&0&0&0&0&0&0&0&0\\ 0&0&1&0&0&0&0&0&0&0\\ 0&0&0&1&0&0&0&0&0&0\\ 0&0&0&0&1&0&0&0&0&0\\ 0&0&0&0&0&1&0&0&0&0\\ 0&0&0&0&0&0&1&0&0&0\\ 0&0&0&0&0&0&0&1&0&0\\ 0&0&0&0&0&0&0&0&1&0\\ 0&0&0&0&0&0&0&0&0&1 \end{bmatrix} \quad \bar{\nu}_e = \begin{bmatrix} -1&0&0&0&0&0&0&0&0&0\\ 0&-1&0&0&0&0&0&0&0&0\\ 0&0&-1&0&0&0&0&0&0&0\\ 0&0&0&-1&0&0&0&0&0&0\\ 0&0&0&0& &0&0&0&0&0\\ 0&0&0&0&0&1&0&0&0&0\\ 0&0&0&0&0&0&1&0&0&0\\ 0&0&0&0&0&0&0&1&0&0\\ 0&0&0&0&0&0&0&0&1&0\\ 0&0&0&0&0&0&0&0&0&1 \end{bmatrix} \qquad (4.9)$$

$$\nu_\mu = \begin{bmatrix} 0&1&0&0&0&0&0&0&0&0\\ &0&1&0&0&0&0&0&0&0\\ 1&0&0&0&0&0&0&0&0&0\\ 0&0&0&1&0&0&0&0&0&0\\ 0&0&0&0&1&0&0&0&0&0\\ 0&0&0&0&0&1&0&0&0&0\\ 0&0&0&0&0&0&1&0&0&0\\ 0&0&0&0&0&0&0&0&1&0\\ 0&0&0&0&0&0&0&0&0&1\\ 0&0&0&0&0&0&0&1&0&0 \end{bmatrix} \quad \bar{\nu}_\mu = \begin{bmatrix} 0&-1&0&0&0&0&0&0&0&0\\ 0&0&-1&0&0&0&0&0&0&0\\ -1&0&0&0&0&0&0&0&0&0\\ 0&0&0&-1&0&0&0&0&0&0\\ 0&0&0&0&1&0&0&0&0&0\\ 0&0&0&0&0&1&0&0&0&0\\ 0&0&0&0&0&0&1&0&0&0\\ 0&0&0&0&0&0&0&0&1&0\\ 0&0&0&0&0&0&0&0&0&1\\ 0&0&0&0&0&0&0&1&0&0 \end{bmatrix}$$

$$\nu_\tau = \begin{bmatrix} 0&0&1&0&0&0&0&0&0&0\\ 1&0&0&0&0&0&0&0&0&0\\ 0&1&0&0&0&0&0&0&0&0\\ 0&0&0&1&0&0&0&0&0&0\\ 0&0&0&0&1&0&0&0&0&0\\ 0&0&0&0&0&1&0&0&0&0\\ 0&0&0&0&0&0&1&0&0&0\\ 0&0&0&0&0&0&0&0&0&1\\ 0&0&0&0&0&0&0&1&0&0\\ 0&0&0&0&0&0&0&0&1&0 \end{bmatrix} \quad \bar{\nu}_\tau = \begin{bmatrix} 0&0&-1&0&0&0&0&0&0&0\\ -1&0&0&0&0&0&0&0&0&0\\ 0&-1&0&0&0&0&0&0&0&0\\ 0&0&0&-1&0&0&0&0&0&0\\ 0&0&0&0&1&0&0&0&0&0\\ 0&0&0&0&0&1&0&0&0&0\\ 0&0&0&0&0&0&1&0&0&0\\ 0&0&0&0&0&0&0&0&0&1\\ 0&0&0&0&0&0&0&1&0&0\\ 0&0&0&0&0&0&0&0&1&0 \end{bmatrix}$$

$$e = \begin{bmatrix} -1 & 0 & 0 & 0 & 0 & 0 & 0 & 0 & 0 & 0 \\ 0 & -1 & 0 & 0 & 0 & 0 & 0 & 0 & 0 & 0 \\ 0 & 0 & -1 & 0 & 0 & 0 & 0 & 0 & 0 & 0 \\ 0 & 0 & 0 & i & 0 & 0 & 0 & 0 & 0 & 0 \\ 0 & 0 & 0 & 0 & 1 & 0 & 0 & 0 & 0 & 0 \\ 0 & 0 & 0 & 0 & 0 & 1 & 0 & 0 & 0 & 0 \\ 0 & 0 & 0 & 0 & 0 & 0 & 1 & 0 & 0 & 0 \\ 0 & 0 & 0 & 0 & 0 & 0 & 0 & 1 & 0 & 0 \\ 0 & 0 & 0 & 0 & 0 & 0 & 0 & 0 & 1 & 0 \\ 0 & 0 & 0 & 0 & 0 & 0 & 0 & 0 & 0 & 1 \end{bmatrix} \qquad \bar{e} = \begin{bmatrix} 1 & 0 & 0 & 0 & 0 & 0 & 0 & 0 & 0 & 0 \\ 0 & 1 & 0 & 0 & 0 & 0 & 0 & 0 & 0 & 0 \\ 0 & 0 & 1 & 0 & 0 & 0 & 0 & 0 & & 0 \\ 0 & 0 & 0 & -i & 0 & 0 & 0 & 0 & 0 & 0 \\ 0 & 0 & 0 & 0 & 1 & 0 & 0 & 0 & 0 & 0 \\ 0 & 0 & 0 & 0 & 0 & 1 & 0 & 0 & 0 & 0 \\ 0 & 0 & 0 & 0 & 0 & 0 & 1 & 0 & 0 & 0 \\ 0 & 0 & 0 & 0 & 0 & 0 & 0 & 1 & 0 & 0 \\ 0 & 0 & 0 & 0 & 0 & 0 & 0 & 0 & 1 & 0 \\ 0 & 0 & 0 & 0 & 0 & 0 & 0 & 0 & 0 & 1 \end{bmatrix}$$

$$\mu = \begin{bmatrix} 0 & -1 & 0 & 0 & 0 & 0 & 0 & 0 & 0 & 0 \\ 0 & 0 & -1 & 0 & 0 & 0 & 0 & 0 & 0 & 0 \\ -1 & 0 & 0 & 0 & 0 & 0 & 0 & 0 & 0 & 0 \\ 0 & 0 & 0 & i & 0 & 0 & 0 & 0 & 0 & 0 \\ 0 & 0 & 0 & 0 & 0 & 1 & 0 & 0 & 0 & 0 \\ 0 & 0 & 0 & 0 & 0 & 0 & 1 & 0 & 0 & 0 \\ 0 & 0 & 0 & 0 & 1 & 0 & 0 & 0 & 0 & 0 \\ 0 & 0 & 0 & 0 & 0 & 0 & 0 & 1 & 0 & 0 \\ 0 & 0 & 0 & 0 & 0 & 0 & 0 & 0 & 1 & 0 \\ 0 & 0 & 0 & 0 & 0 & 0 & 0 & 0 & 0 & 1 \end{bmatrix} \qquad \bar{\mu} = \begin{bmatrix} 0 & 1 & 0 & 0 & 0 & 0 & 0 & 0 & 0 & 0 \\ 0 & 0 & 1 & 0 & 0 & 0 & 0 & 0 & 0 & 0 \\ 1 & 0 & 0 & 0 & 0 & 0 & 0 & 0 & 0 & 0 \\ 0 & 0 & 0 & -i & 0 & 0 & 0 & 0 & 0 & 0 \\ 0 & 0 & 0 & 0 & 0 & 1 & 0 & 0 & 0 & 0 \\ 0 & 0 & 0 & 0 & 0 & 0 & 1 & 0 & 0 & 0 \\ 0 & 0 & 0 & 0 & 1 & 0 & 0 & 0 & 0 & 0 \\ 0 & 0 & 0 & 0 & 0 & 0 & 0 & 1 & 0 & 0 \\ 0 & 0 & 0 & 0 & 0 & 0 & 0 & 0 & 1 & 0 \\ 0 & 0 & 0 & 0 & 0 & 0 & 0 & 0 & 0 & 1 \end{bmatrix}$$

$$\tau = \begin{bmatrix} 0 & 0 & -1 & 0 & 0 & 0 & 0 & 0 & 0 & 0 \\ -1 & 0 & & 0 & 0 & 0 & 0 & 0 & 0 & 0 \\ 0 & -1 & 0 & 0 & 0 & 0 & 0 & 0 & 0 & 0 \\ 0 & 0 & 0 & i & 0 & 0 & 0 & 0 & 0 & 0 \\ 0 & 0 & 0 & 0 & 0 & 0 & 1 & 0 & 0 & 0 \\ 0 & 0 & 0 & 0 & 1 & 0 & 0 & 0 & 0 & 0 \\ 0 & 0 & 0 & 0 & 0 & 1 & 0 & 0 & 0 & 0 \\ 0 & 0 & 0 & 0 & 0 & 0 & 0 & 1 & 0 & 0 \\ 0 & 0 & 0 & 0 & 0 & 0 & 0 & 0 & 1 & 0 \\ 0 & 0 & 0 & 0 & 0 & 0 & 0 & 0 & 0 & 1 \end{bmatrix} \qquad \bar{\tau} = \begin{bmatrix} 0 & 0 & 1 & 0 & 0 & 0 & 0 & 0 & 0 & 0 \\ 1 & 0 & 0 & 0 & 0 & 0 & 0 & 0 & 0 & 0 \\ 0 & 1 & 0 & 0 & 0 & 0 & 0 & 0 & 0 & 0 \\ 0 & 0 & 0 & -i & 0 & 0 & 0 & 0 & 0 & 0 \\ 0 & 0 & 0 & 0 & 0 & 0 & 1 & 0 & 0 & 0 \\ 0 & 0 & 0 & 0 & 1 & 0 & 0 & 0 & 0 & 0 \\ 0 & 0 & 0 & 0 & 0 & 1 & 0 & 0 & 0 & 0 \\ 0 & 0 & 0 & 0 & 0 & & 0 & 1 & 0 & 0 \\ 0 & 0 & 0 & 0 & 0 & 0 & 0 & 0 & 1 & 0 \\ 0 & 0 & 0 & 0 & 0 & 0 & 0 & 0 & 0 & 1 \end{bmatrix}$$

$$u = \begin{bmatrix} 0 & 1 & 0 & 0 & 0 & 0 & 0 & 0 & 0 & 0 \\ 1 & 0 & 0 & 0 & 0 & 0 & 0 & 0 & 0 & 0 \\ 0 & 0 & 1 & 0 & 0 & 0 & 0 & 0 & 0 & 0 \\ 0 & 0 & 0 & -1 & 0 & 0 & 0 & 0 & 0 & 0 \\ 0 & 0 & 0 & 0 & 1 & 0 & 0 & 0 & 0 & 0 \\ 0 & 0 & 0 & 0 & 0 & 1 & 0 & 0 & 0 & 0 \\ 0 & 0 & 0 & 0 & 0 & 0 & 1 & 0 & 0 & 0 \\ 0 & 0 & 0 & 0 & 0 & 0 & 0 & 0 & 1 & 0 \\ 0 & 0 & 0 & 0 & 0 & 0 & 0 & 1 & 0 & 0 \\ 0 & 0 & 0 & 0 & 0 & 0 & 0 & 0 & 0 & 1 \end{bmatrix} \quad \bar{u} = \begin{bmatrix} 0 & -1 & 0 & 0 & 0 & 0 & 0 & 0 & 0 & 0 \\ -1 & 0 & 0 & 0 & 0 & 0 & 0 & 0 & 0 & 0 \\ 0 & 0 & -1 & 0 & 0 & 0 & 0 & 0 & 0 & 0 \\ 0 & 0 & 0 & 1 & 0 & 0 & 0 & 0 & 0 & 0 \\ 0 & 0 & 0 & 0 & 1 & 0 & 0 & 0 & 0 & 0 \\ 0 & 0 & 0 & 0 & 0 & 1 & 0 & 0 & 0 & 0 \\ 0 & 0 & 0 & 0 & 0 & 0 & 1 & 0 & 0 & 0 \\ 0 & 0 & 0 & 0 & 0 & 0 & 0 & 0 & 1 & 0 \\ 0 & 0 & 0 & 0 & 0 & 0 & 0 & 1 & 0 & 0 \\ 0 & 0 & 0 & 0 & 0 & 0 & 0 & 0 & 0 & 1 \end{bmatrix}$$

$$c = \begin{bmatrix} 0 & 0 & 1 & 0 & 0 & 0 & 0 & 0 & 0 & 0 \\ 0 & 1 & 0 & 0 & 0 & 0 & 0 & 0 & 0 & 0 \\ 1 & 0 & 0 & 0 & 0 & 0 & 0 & 0 & 0 & 0 \\ 0 & 0 & 0 & -1 & 0 & 0 & 0 & 0 & 0 & 0 \\ 0 & 0 & 0 & 0 & 1 & 0 & 0 & 0 & 0 & 0 \\ 0 & 0 & 0 & 0 & 0 & 1 & 0 & 0 & 0 & 0 \\ 0 & 0 & 0 & 0 & 0 & 0 & 1 & 0 & 0 & 0 \\ 0 & 0 & 0 & 0 & 0 & 0 & 0 & 0 & 0 & 1 \\ 0 & 0 & 0 & 0 & 0 & 0 & 0 & 0 & 1 & 0 \\ 0 & 0 & 0 & 0 & 0 & 0 & 0 & 1 & 0 & 0 \end{bmatrix} \quad \bar{c} = \begin{bmatrix} 0 & 0 & -1 & 0 & 0 & 0 & 0 & 0 & 0 & 0 \\ 0 & -1 & 0 & 0 & 0 & 0 & 0 & 0 & 0 & 0 \\ -1 & 0 & 0 & 0 & 0 & 0 & 0 & 0 & 0 & 0 \\ 0 & 0 & 0 & 1 & 0 & 0 & 0 & 0 & 0 & 0 \\ 0 & 0 & 0 & 0 & 1 & 0 & 0 & 0 & 0 & 0 \\ 0 & 0 & 0 & 0 & 0 & 1 & 0 & 0 & 0 & 0 \\ 0 & 0 & 0 & 0 & 0 & 0 & 1 & 0 & 0 & 0 \\ 0 & 0 & 0 & 0 & 0 & 0 & 0 & 0 & 0 & 1 \\ 0 & 0 & 0 & 0 & 0 & 0 & 0 & 0 & 1 & 0 \\ 0 & 0 & 0 & 0 & 0 & 0 & 0 & 1 & 0 & 0 \end{bmatrix}$$

$$t = \begin{bmatrix} 1 & 0 & 0 & 0 & 0 & 0 & 0 & 0 & 0 & 0 \\ 0 & 0 & 1 & 0 & 0 & 0 & 0 & 0 & 0 & 0 \\ 0 & 1 & 0 & 0 & 0 & 0 & 0 & 0 & 0 & 0 \\ 0 & 0 & 0 & -1 & 0 & 0 & 0 & 0 & 0 & 0 \\ 0 & 0 & 0 & 0 & 1 & 0 & 0 & 0 & 0 & 0 \\ 0 & 0 & 0 & 0 & 0 & 1 & 0 & 0 & 0 & 0 \\ 0 & 0 & 0 & 0 & 0 & 0 & 1 & 0 & 0 & 0 \\ 0 & 0 & 0 & 0 & 0 & 0 & 0 & 1 & 0 & 0 \\ 0 & 0 & 0 & 0 & 0 & 0 & 0 & 0 & 0 & 1 \\ 0 & 0 & 0 & 0 & 0 & 0 & 0 & 0 & 1 & 0 \end{bmatrix} \quad \bar{t} = \begin{bmatrix} -1 & 0 & 0 & 0 & 0 & 0 & 0 & 0 & 0 & 0 \\ 0 & 0 & -1 & 0 & 0 & 0 & 0 & 0 & 0 & 0 \\ 0 & -1 & 0 & 0 & 0 & 0 & 0 & 0 & 0 & 0 \\ 0 & 0 & 0 & 1 & 0 & 0 & 0 & 0 & 0 & 0 \\ 0 & 0 & 0 & 0 & 1 & 0 & 0 & 0 & 0 & 0 \\ 0 & 0 & 0 & 0 & 0 & 1 & 0 & 0 & 0 & 0 \\ 0 & 0 & 0 & 0 & 0 & 0 & 1 & 0 & 0 & 0 \\ 0 & 0 & 0 & 0 & 0 & 0 & 0 & 1 & 0 & 0 \\ 0 & 0 & 0 & 0 & 0 & 0 & 0 & 0 & 0 & 1 \\ 0 & 0 & 0 & 0 & 0 & 0 & 0 & 0 & 1 & 0 \end{bmatrix}$$

$$d=\begin{bmatrix} 0 & -1 & 0 & 0 & 0 & 0 & 0 & 0 & 0 & 0 \\ -1 & 0 & 0 & 0 & 0 & 0 & 0 & 0 & 0 & 0 \\ 0 & 0 & -1 & 0 & 0 & 0 & 0 & 0 & 0 & 0 \\ 0 & 0 & 0 & -i & 0 & 0 & 0 & 0 & 0 & 0 \\ 0 & 0 & 0 & 0 & 0 & 1 & 0 & 0 & 0 & 0 \\ 0 & 0 & 0 & 0 & 1 & 0 & 0 & 0 & 0 & 0 \\ 0 & 0 & 0 & 0 & 0 & 0 & 1 & 0 & 0 & 0 \\ 0 & 0 & 0 & 0 & 0 & 0 & 0 & 1 & 0 & 0 \\ 0 & 0 & 0 & 0 & 0 & 0 & 0 & 0 & 1 & 0 \\ 0 & 0 & 0 & 0 & 0 & 0 & 0 & 0 & 0 & 1 \end{bmatrix} \qquad \bar{d}=\begin{bmatrix} 0 & 1 & 0 & 0 & 0 & 0 & 0 & 0 & 0 & 0 \\ 1 & 0 & 0 & 0 & 0 & 0 & 0 & 0 & 0 & 0 \\ 0 & 0 & 1 & 0 & 0 & 0 & 0 & 0 & 0 & 0 \\ 0 & 0 & 0 & i & 0 & 0 & 0 & 0 & 0 & 0 \\ 0 & 0 & 0 & 0 & 0 & 1 & 0 & 0 & 0 & 0 \\ 0 & 0 & 0 & 0 & 1 & 0 & 0 & 0 & 0 & 0 \\ 0 & 0 & 0 & 0 & 0 & 0 & 1 & 0 & 0 & 0 \\ 0 & 0 & 0 & 0 & 0 & 0 & 0 & 1 & 0 & 0 \\ 0 & 0 & 0 & 0 & 0 & 0 & 0 & 0 & 1 & 0 \\ 0 & 0 & 0 & 0 & 0 & 0 & 0 & 0 & 0 & 1 \end{bmatrix}$$

$$s=\begin{bmatrix} 0 & 0 & -1 & 0 & 0 & 0 & 0 & 0 & 0 & 0 \\ 0 & -1 & 0 & 0 & 0 & 0 & 0 & 0 & 0 & 0 \\ -1 & 0 & 0 & 0 & 0 & 0 & 0 & 0 & 0 & 0 \\ 0 & 0 & 0 & -i & 0 & 0 & 0 & 0 & 0 & 0 \\ 0 & 0 & 0 & 0 & 0 & 0 & 1 & 0 & 0 & 0 \\ 0 & 0 & 0 & 0 & 0 & 1 & 0 & 0 & 0 & 0 \\ 0 & 0 & 0 & 0 & 1 & 0 & 0 & 0 & 0 & 0 \\ 0 & 0 & 0 & 0 & 0 & 0 & 0 & 1 & 0 & 0 \\ 0 & 0 & 0 & 0 & 0 & 0 & 0 & 0 & 1 & 0 \\ 0 & 0 & 0 & 0 & 0 & 0 & 0 & 0 & 0 & 1 \end{bmatrix} \qquad \bar{s}=\begin{bmatrix} 0 & 0 & 1 & 0 & 0 & 0 & 0 & 0 & 0 & 0 \\ 0 & 1 & 0 & 0 & 0 & 0 & 0 & 0 & 0 & 0 \\ 1 & 0 & 0 & 0 & 0 & 0 & 0 & 0 & 0 & 0 \\ 0 & 0 & 0 & i & 0 & 0 & 0 & 0 & 0 & 0 \\ 0 & 0 & 0 & 0 & 0 & 0 & 1 & 0 & 0 & 0 \\ 0 & 0 & 0 & 0 & 0 & 1 & 0 & 0 & 0 & 0 \\ 0 & 0 & 0 & 0 & 1 & 0 & 0 & 0 & 0 & 0 \\ 0 & 0 & 0 & 0 & 0 & 0 & 0 & 1 & 0 & 0 \\ 0 & 0 & 0 & 0 & 0 & 0 & 0 & 0 & 1 & 0 \\ 0 & 0 & 0 & 0 & 0 & 0 & 0 & 0 & 0 & 1 \end{bmatrix}$$

$$b=\begin{bmatrix} -1 & 0 & 0 & 0 & 0 & 0 & 0 & 0 & 0 & 0 \\ 0 & 0 & -1 & 0 & 0 & 0 & 0 & 0 & 0 & 0 \\ 0 & -1 & 0 & 0 & 0 & 0 & 0 & 0 & 0 & 0 \\ 0 & 0 & 0 & -i & 0 & 0 & 0 & 0 & 0 & 0 \\ 0 & 0 & 0 & 0 & 1 & 0 & 0 & 0 & 0 & 0 \\ 0 & 0 & 0 & 0 & 0 & 0 & 1 & 0 & 0 & 0 \\ 0 & 0 & 0 & 0 & 0 & 1 & 0 & 0 & 0 & 0 \\ 0 & 0 & 0 & 0 & 0 & 0 & 0 & 1 & 0 & 0 \\ 0 & 0 & 0 & 0 & 0 & 0 & 0 & 0 & 1 & 0 \\ 0 & 0 & 0 & 0 & 0 & 0 & 0 & 0 & 0 & 1 \end{bmatrix} \qquad \bar{b}=\begin{bmatrix} 1 & 0 & 0 & 0 & 0 & 0 & 0 & 0 & 0 & 0 \\ 0 & 0 & 1 & 0 & 0 & 0 & 0 & 0 & 0 & 0 \\ 0 & 1 & 0 & 0 & 0 & 0 & 0 & 0 & 0 & 0 \\ 0 & 0 & 0 & i & 0 & 0 & 0 & 0 & 0 & 0 \\ 0 & 0 & 0 & 0 & 1 & 0 & 0 & 0 & 0 & 0 \\ 0 & 0 & 0 & 0 & 0 & 0 & 1 & 0 & 0 & 0 \\ 0 & 0 & 0 & 0 & 0 & 1 & 0 & 0 & 0 & 0 \\ 0 & 0 & 0 & 0 & 0 & 0 & 0 & 1 & 0 & 0 \\ 0 & 0 & 0 & 0 & 0 & 0 & 0 & 0 & 1 & 0 \\ 0 & 0 & 0 & 0 & 0 & 0 & 0 & 0 & 0 & 1 \end{bmatrix}$$

Let us write the ten generators of the Poincaré group $\mathbf{I}^{(3,\,1)}$ as a column vector

$$\begin{bmatrix} P_X \\ P_Y \\ P_Z \\ P_T \\ J_X \\ J_Y \\ J_Z \\ N_X \\ N_Y \\ N_Z \end{bmatrix} \tag{4.10}$$

Then each of the twenty-four fermions given as 10×10 matrices in (4.9) act on the column vector (4.10) by left multiplication. Note that each fermion F will permute the infinitesimal space translations $\{\pm P_X, \pm P_Y, \pm P_Z\}$ amongst themselves, multiply the infinitesimal time translation P_T by ± 1 or $\pm i$, permute the infinitesimal Lorentz boosts $\{N_X, N_Y, N_Z\}$ amongst themselves and permute the infinitesimal components of angular momentum $\{J_X, J_Y, J_Z\}$ amongst themselves. Also, each of the twenty-four fermions given as 10×10 matrices in (4.9) are unitary matrices. Thus, each fermion F induces an isomorphic copy of the complex group algebra of the Poincaré group $\mathbf{CI}^{(3,\,1)}$. The irreducible unitary representation of the fermion F according to Wigner's classification is equivalent to a simple module M over this isomorphic copy of the complex group algebra of the Poincaré group $\mathbf{CI}^{(3,\,1)}$. Thus, the twenty-four 10×10 matrices in (4.9) label the twenty-four unitary irreducible representations for the fermions given by Wigner's classification. Similarly, by the boson selection rule in **[2]**, each spin 1 boson is given by four 10×10 unitary matrices and the unique spin 0 Higgs boson is given by twenty-four 10×10 unitary matrices. Since the grand unified theory **[2]** already specifies the spin, charge and mass for each particle of the standard model there is a perfect match with their unitary irreducible representations according to Wigner's classification.

Finally, we define the *Steiner System of Fermions* $S(5, 8, 24)$ according to the proof of the four color theorem **[1]**: there are 24 fermions, there are 759 blocks with 8 fermions in each block such that any set of 5 fermions is contained in a unique block (see the appedix). The following program for Microsoft Windows generates the Steiner system of fermions explicitly:

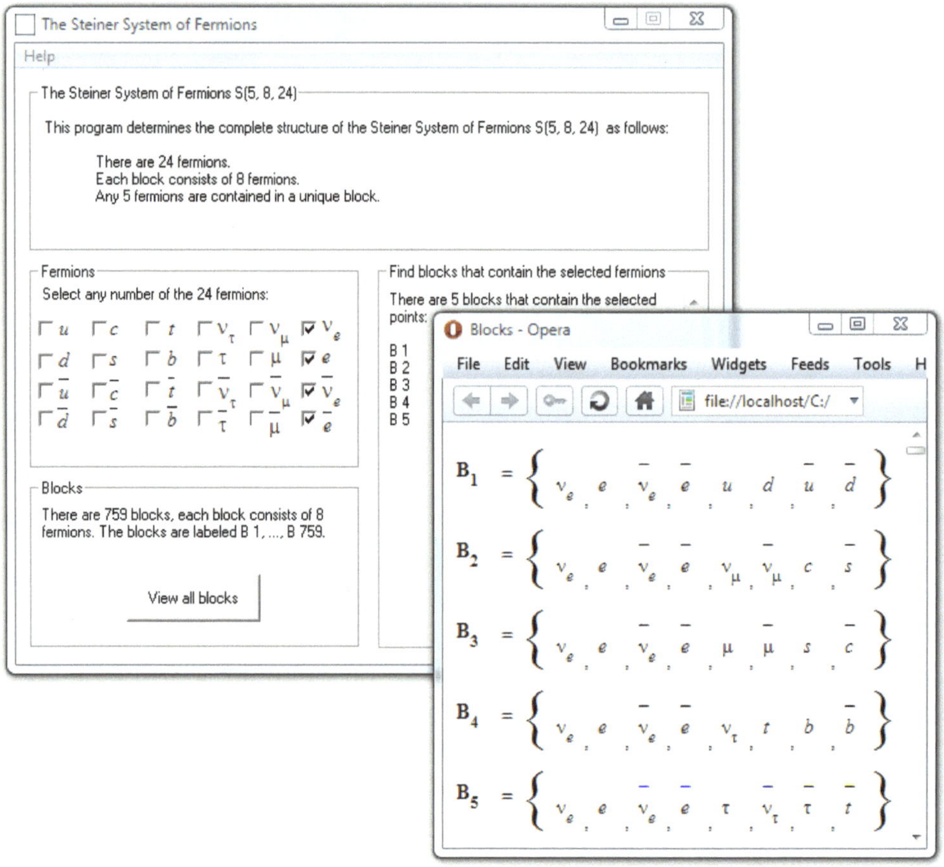

Program 4.1. *The Steiner System of Fermions*
http://www.dharwadker.org/space_time

The automorphism group of $S(5, 8, 24)$ is the famous 5-transitive Mathieu group $\mathbf{M}_{24}$ on 24 letters **[21][22][23]**. Thus, the Mathieu group $\mathbf{M}_{24}$ acts as a group of symmetries of the 24 fermions preserving the structure of the blocks of 8 fermions. Note that the three generations of fermions appear naturally as blocks of the Steiner system:

$$
\begin{array}{llcll}
\text{Generation I} & : \mathbf{B_1} & = & \{\ u, d, \bar{u}, \bar{d}, \nu_e, e, \bar{\nu_e}, \bar{e}\ \} & \\
\text{Generation II} & : \mathbf{B_{730}} & = & \{\ c, s, \bar{c}, \bar{s}, \nu_\mu, \mu, \bar{\nu_\mu}, \bar{\mu}\ \} & (4.11) \\
\text{Generation III} & : \mathbf{B_{759}} & = & \{\ t, b, \bar{t}, \bar{b}, \nu_\tau, \tau, \bar{\nu_\tau}, \bar{\tau}\ \} &
\end{array}
$$

The pattern (1, 3, 8, 24) of the generators of the grand unification sequence $\boldsymbol{U}(1) \rightarrow \boldsymbol{SU}(2) \rightarrow \boldsymbol{SU}(3) \rightarrow \boldsymbol{SU}(5)$ reappears in many contexts. Starting from the 24 fermions, we select the blocks of 8 fermions to form the Steiner system. From all the blocks of the Steiner system, we select the 3 generations of fermions (4.11). Finally, from the 3 generations of fermions we may select 1 particular generation. For example, the first generation is exactly the set of fermions that participate in beta-decay. We conjecture that all observable interactions of particles obeying the various conservation laws of physics **[24]** can be found among the blocks of the Steiner system of fermions.

Appendix: The Blocks of the Steiner System of Fermions

$$\mathbf{B}_1 = \left\{ \nu_e, e, \bar{\nu}_e, \bar{e}, u, d, \bar{u}, \bar{d} \right\}$$

$$\mathbf{B}_2 = \left\{ \nu_e, e, \bar{\nu}_e, \bar{e}, \nu_\mu, \bar{\nu}_\mu, c, \bar{s} \right\}$$

$$\mathbf{B}_3 = \left\{ \nu_e, e, \bar{\nu}_e, \bar{e}, \mu, \bar{\mu}, s, \bar{c} \right\}$$

$$\mathbf{B}_4 = \left\{ \nu_e, e, \bar{\nu}_e, \bar{e}, \nu_\tau, t, b, \bar{b} \right\}$$

$$\mathbf{B}_5 = \left\{ \nu_e, e, \bar{\nu}_e, \bar{e}, \tau, \bar{\nu}_\tau, \bar{\tau}, \bar{t} \right\}$$

$$\mathbf{B}_6 = \left\{ \nu_e, e, \bar{\nu}_e, u, \nu_\mu, \bar{c}, \bar{\nu}_\tau, b \right\}$$

$$\mathbf{B}_7 = \left\{ \nu_e, e, \bar{\nu}_e, u, \mu, \bar{\nu}_\mu, \nu_\tau, \bar{\tau} \right\}$$

$$\mathbf{B}_8 = \left\{ \nu_e, e, \bar{\nu}_e, u, \bar{\mu}, \bar{s}, \tau, \bar{b} \right\}$$

$$\mathbf{B}_9 = \left\{ \nu_e, e, \bar{\nu}_e, u, c, s, t, \bar{t} \right\}$$

$$\mathbf{B}_{10} = \left\{ \nu_e, e, \bar{\nu}_e, d, \nu_\mu, s, \nu_\tau, \tau \right\}$$

$$\mathbf{B}_{11} = \left\{ \nu_e, e, \bar{\nu}_e, d, \mu, \bar{s}, \bar{\nu}_\tau, t \right\}$$

$$\mathbf{B}_{12} = \left\{ \nu_e, e, \bar{\nu}_e, d, \bar{\nu}_\mu, \bar{\mu}, b, \bar{t} \right\}$$

$$\mathbf{B}_{13} = \left\{ \nu_e, e, \bar{\nu}_e, d, c, \bar{c}, \bar{\tau}, \bar{b} \right\}$$

$$\mathbf{B}_{14} = \left\{ \nu_e, e, \bar{\nu}_e, \bar{u}, \nu_\mu, \mu, \bar{t}, \bar{b} \right\}$$

$$\mathbf{B}_{15} = \left\{ \nu_e, e, \bar{\nu}_e, \bar{u}, \bar{\nu}_\mu, \bar{c}, \tau, t \right\}$$

$$\mathbf{B}_{16} = \left\{ \nu_e, e, \bar{\nu}_e, \bar{u}, \bar{\mu}, c, \nu_\tau, \bar{\nu}_\tau \right\}$$

$$\mathbf{B}_{17} = \left\{ \nu_e, e, \bar{\nu}_e, \bar{u}, s, \bar{s}, \bar{\tau}, b \right\}$$

$$\mathbf{B}_{18} = \left\{ \nu_e, e, \bar{\nu}_e, \bar{d}, \nu_\mu, \bar{\mu}, \bar{\tau}, t \right\}$$

$$\mathbf{B}_{19} = \left\{ \nu_e, e, \bar{\nu}_e, \bar{d}, \mu, c, \tau, b \right\}$$

$$\mathbf{B}_{20} = \left\{ \nu_e, e, \bar{\nu}_e, \bar{d}, \bar{\nu}_\mu, s, \bar{\nu}_\tau, \bar{b} \right\}$$

$$\mathbf{B}_{21} = \left\{ \nu_e, e, \bar{\nu}_e, \bar{d}, \bar{c}, \bar{s}, \nu_\tau, \bar{t} \right\}$$

$$\mathbf{B}_{22} = \left\{ \nu_e, e, \bar{e}, u, \nu_\mu, s, \bar{\tau}, \bar{b} \right\}$$

$$\mathbf{B}_{23} = \left\{ \nu_e, e, \bar{e}, u, \mu, \bar{s}, b, \bar{t} \right\}$$

$$\mathbf{B}_{24} = \left\{ \nu_e, e, \bar{e}, u, \bar{\nu}_\mu, \bar{\mu}, \bar{\nu}_\tau, t \right\}$$

$$\mathbf{B}_{25} = \left\{ \nu_e, e, \bar{e}, u, c, \bar{c}, \nu_\tau, \tau \right\}$$

$$\mathbf{B}_{26} = \left\{ \nu_e, e, \bar{e}, d, \nu_\mu, \bar{c}, t, \bar{t} \right\}$$

$$\mathbf{B}_{27} = \left\{ \nu_e, e, \bar{e}, d, \mu, \bar{\nu}_\mu, \tau, \bar{b} \right\}$$

$$\mathbf{B}_{28} = \left\{ \nu_e, e, \bar{e}, d, \bar{\mu}, \bar{s}, \nu_\tau, \bar{\tau} \right\}$$

$$\mathbf{B}_{29} = \left\{ \nu_e, e, \bar{e}, d, c, s, \bar{\nu}_\tau, b \right\}$$

$$\mathbf{B}_{30} = \left\{ \nu_e, e, \bar{e}, \bar{u}, \nu_\mu, \bar{\mu}, \tau, b \right\}$$

$$\mathbf{B}_{31} = \left\{ \nu_e, e, \bar{e}, \bar{u}, \mu, c, \bar{\tau}, t \right\}$$

$$\mathbf{B}_{32} = \left\{ \nu_e, e, \bar{e}, \bar{u}, \bar{\nu}_\mu, s, \nu_\tau, \bar{t} \right\}$$

$$\mathbf{B}_{33} = \left\{ \nu_e, e, \bar{e}, \bar{u}, \bar{c}, \bar{s}, \bar{\nu}_\tau, \bar{b} \right\}$$

$$\mathbf{B}_{34} = \left\{ \nu_e, e, \bar{e}, \bar{d}, \nu_\mu, \mu, \nu_\tau, \bar{\nu}_\tau \right\}$$

$$\mathbf{B}_{35} = \left\{ \nu_e, e, \bar{e}, \bar{d}, \bar{\nu}_\mu, \bar{c}, \bar{\tau}, b \right\}$$

$$\mathbf{B}_{36} = \left\{ \nu_e, e, \bar{e}, \bar{d}, \bar{\mu}, c, \bar{t}, \bar{b} \right\}$$

$$\mathbf{B}_{37} = \left\{ \nu_e, e, \bar{e}, \bar{d}, s, \bar{s}, \tau, t \right\}$$

$$\mathbf{B}_{38} = \left\{ \nu_e, e, u, d, \nu_\mu, \mu, \bar{\mu}, c \right\}$$

$$\mathbf{B}_{39} = \left\{ \nu_e, e, u, d, \bar{\nu}_\mu, s, \bar{c}, \bar{s} \right\}$$

$$\mathbf{B}_{40} = \left\{ \nu_e, e, u, d, \nu_\tau, \bar{\nu}_\tau, \bar{t}, \bar{b} \right\}$$

$$\mathbf{B}_{41} = \left\{ \nu_e, e, u, d, \tau, \bar{\tau}, t, b \right\}$$

$$\mathbf{B}_{42} = \left\{ \nu_e, e, u, \bar{u}, \nu_\mu, \bar{s}, \nu_\tau, t \right\}$$

$$\mathbf{B}_{43} = \left\{ \nu_e, e, u, \bar{u}, \mu, s, \tau, \bar{\nu}_\tau \right\}$$

$$\mathbf{B}_{44} = \left\{ \nu_e, e, u, \bar{u}, \bar{\nu}_\mu, c, b, \bar{b} \right\}$$

$$\mathbf{B}_{45} = \left\{ \nu_e, e, u, \bar{u}, \bar{\mu}, \bar{c}, \bar{\tau}, \bar{t} \right\}$$

$$\mathbf{B}_{46} = \left\{ \nu_e, e, u, \bar{d}, \nu_\mu, \bar{\nu}_\mu, \tau, \bar{t} \right\}$$

$$\mathbf{B}_{47} = \left\{ \nu_e, e, u, \bar{d}, \mu, \bar{c}, t, \bar{b} \right\}$$

$$\mathbf{B}_{48} = \left\{ \nu_e, e, u, \bar{d}, \bar{\mu}, s, \nu_\tau, b \right\}$$

$$\mathbf{B}_{49} = \left\{ \nu_e, e, u, \bar{d}, c, \bar{s}, \bar{\nu}_\tau, \bar{\tau} \right\}$$

$$\mathbf{B}_{50} = \left\{ \nu_e, e, d, \bar{u}, \nu_\mu, \bar{\nu}_\mu, \bar{\nu}_\tau, \bar{\tau} \right\}$$

$$\mathbf{B}_{51} = \left\{ \nu_e, e, d, \bar{u}, \mu, \bar{c}, \nu_\tau, b \right\}$$

$$\mathbf{B}_{52} = \left\{ \nu_e, e, d, \bar{u}, \bar{\mu}, s, t, \bar{b} \right\}$$

$$\mathbf{B}_{53} = \left\{ \nu_e, e, d, \bar{u}, c, \bar{s}, \tau, \bar{t} \right\}$$

$$\mathbf{B}_{54} = \left\{ \nu_e, e, d, \bar{d}, \nu_\mu, \bar{s}, b, \bar{b} \right\}$$

$$\mathbf{B}_{55} = \left\{ \nu_e, e, d, \bar{d}, \mu, s, \bar{\tau}, \bar{t} \right\}$$

$$\mathbf{B}_{56} = \left\{ \nu_e, e, d, \bar{d}, \bar{\nu}_\mu, c, \nu_\tau, t \right\}$$

$$\mathbf{B}_{57} = \left\{ \nu_e, e, d, \bar{d}, \bar{\mu}, \bar{c}, \tau, \bar{\nu}_\tau \right\}$$

$$\mathbf{B}_{58} = \left\{ \nu_e, e, \bar{u}, \bar{d}, \nu_\mu, c, s, \bar{c} \right\}$$

$$\mathbf{B}_{59} = \left\{ \nu_e, e, \bar{u}, \bar{d}, \mu, \bar{\nu}_\mu, \bar{\mu}, \bar{s} \right\}$$

$$\mathbf{B}_{60} = \left\{ \nu_e, e, \bar{u}, \bar{d}, \nu_\tau, \tau, \bar{\tau}, \bar{b} \right\}$$

$$\mathbf{B}_{61} = \left\{ \nu_e, e, \bar{u}, \bar{d}, \bar{\nu}_\tau, t, b, \bar{t} \right\}$$

$$\mathbf{B}_{62} = \left\{ \nu_e, e, \nu_\mu, \mu, \bar{\nu}_\mu, s, t, b \right\}$$

$$\mathbf{B}_{63} = \left\{ \nu_e, e, \nu_\mu, \mu, \bar{c}, \bar{s}, \tau, \bar{\tau} \right\}$$

$$\mathbf{B}_{64} = \left\{ \nu_e, e, \nu_\mu, \bar{\nu}_\mu, \bar{\mu}, \bar{c}, \nu_\tau, \bar{b} \right\}$$

$$\mathbf{B}_{65} = \left\{ \nu_e, e, \nu_\mu, \bar{\mu}, s, \bar{s}, \bar{\nu}_\tau, \bar{t} \right\}$$

$$\mathbf{B}_{66} = \left\{ \nu_e, e, \nu_\mu, c, \nu_\tau, \bar{\tau}, b, \bar{t} \right\}$$

$$\mathbf{B}_{67} = \left\{ \nu_e, e, \nu_\mu, c, \tau, \bar{\nu}_\tau, t, \bar{b} \right\}$$

$$\mathbf{B}_{68} = \left\{ \nu_e, e, \mu, \bar{\nu}_\mu, c, \bar{c}, \bar{\nu}_\tau, \bar{t} \right\}$$

$$\mathbf{B}_{69} = \left\{ \nu_e, e, \mu, \bar{\mu}, \nu_\tau, \tau, t, \bar{t} \right\}$$

$$\mathbf{B}_{70} = \left\{ \nu_e, e, \mu, \bar{\mu}, \bar{\nu}_\tau, \bar{\tau}, b, \bar{b} \right\}$$

$$\mathbf{B}_{71} = \left\{ \nu_e, e, \mu, c, s, \bar{s}, \nu_\tau, \bar{b} \right\}$$

$$\mathbf{B}_{72} = \left\{ \nu_e, e, \bar{\nu}_\mu, \bar{\mu}, c, s, \tau, \bar{\tau} \right\}$$

$$\mathbf{B}_{73} = \left\{ \nu_e, e, \bar{\nu}_\mu, \bar{s}, \nu_\tau, \tau, \bar{\nu}_\tau, b \right\}$$

$$\mathbf{B}_{74} = \left\{ \nu_e, e, \bar{\nu}_\mu, \bar{s}, \bar{\tau}, t, \bar{t}, \bar{b} \right\}$$

$$\mathbf{B}_{75} = \left\{ \nu_e, e, \bar{\mu}, c, \bar{c}, \bar{s}, t, b \right\}$$

$$\mathbf{B}_{76} = \left\{ \nu_e, e, s, \bar{c}, \nu_\tau, \bar{\nu}_\tau, \bar{\tau}, t \right\}$$

$$\mathbf{B}_{77} = \left\{ \nu_e, e, s, \bar{c}, \tau, b, \bar{t}, \bar{b} \right\}$$

$$\mathbf{B}_{78} = \left\{ \nu_e, \bar{\nu}_e, \bar{e}, u, \nu_\mu, \bar{\mu}, \nu_\tau, \bar{t} \right\}$$

$$\mathbf{B}_{79} = \left\{ \nu_e, \bar{\nu}_e, \bar{e}, u, \mu, c, \bar{\nu}_\tau, \bar{b} \right\}$$

$$\mathbf{B}_{80} = \left\{ \nu_e, \bar{\nu}_e, \bar{e}, u, \bar{\nu}_\mu, s, \tau, b \right\}$$

$$\mathbf{B}_{81} = \left\{ \nu_e, \bar{\nu}_e, \bar{e}, u, \bar{c}, \bar{s}, \bar{\tau}, t \right\}$$

$$\mathbf{B}_{82} = \left\{ \nu_e, \bar{\nu}_e, \bar{e}, d, \nu_\mu, \mu, \bar{\tau}, b \right\}$$

$$\mathbf{B}_{83} = \left\{ \nu_e, \bar{\nu}_e, \bar{e}, d, \bar{\nu}_\mu, \bar{c}, \nu_\tau, \bar{\nu}_\tau \right\}$$

$$\mathbf{B_{84}} = \left\{ \nu_e, \bar{\nu}_e, \bar{e}, d, \bar{\mu}, c, \tau, t \right\}$$

$$\mathbf{B_{85}} = \left\{ \nu_e, \bar{\nu}_e, \bar{e}, d, s, \bar{s}, \bar{t}, \bar{b} \right\}$$

$$\mathbf{B_{86}} = \left\{ \nu_e, \bar{\nu}_e, \bar{e}, \bar{u}, \nu_\mu, s, \bar{\nu}_\tau, t \right\}$$

$$\mathbf{B_{87}} = \left\{ \nu_e, \bar{\nu}_e, \bar{e}, \bar{u}, \mu, \bar{s}, \nu_\tau, \tau \right\}$$

$$\mathbf{B_{88}} = \left\{ \nu_e, \bar{\nu}_e, \bar{e}, \bar{u}, \bar{\nu}_\mu, \bar{\mu}, \bar{\tau}, \bar{b} \right\}$$

$$\mathbf{B_{89}} = \left\{ \nu_e, \bar{\nu}_e, \bar{e}, \bar{u}, c, \bar{c}, b, \bar{t} \right\}$$

$$\mathbf{B_{90}} = \left\{ \nu_e, \bar{\nu}_e, \bar{e}, \bar{d}, \nu_\mu, \bar{c}, \tau, \bar{b} \right\}$$

$$\mathbf{B_{91}} = \left\{ \nu_e, \bar{\nu}_e, \bar{e}, \bar{d}, \mu, \bar{\nu}_\mu, t, \bar{t} \right\}$$

$$\mathbf{B_{92}} = \left\{ \nu_e, \bar{\nu}_e, \bar{e}, \bar{d}, \bar{\mu}, \bar{s}, \bar{\nu}_\tau, b \right\}$$

$$\mathbf{B_{93}} = \left\{ \nu_e, \bar{\nu}_e, \bar{e}, \bar{d}, c, s, \nu_\tau, \bar{\tau} \right\}$$

$$\mathbf{B_{94}} = \left\{ \nu_e, \bar{\nu}_e, u, d, \nu_\mu, \bar{\nu}_\mu, t, \bar{b} \right\}$$

$$\mathbf{B_{95}} = \left\{ \nu_e, \bar{\nu}_e, u, d, \mu, \bar{c}, \tau, \bar{t} \right\}$$

$$\mathbf{B_{96}} = \left\{ \nu_e, \bar{\nu}_e, u, d, \bar{\mu}, s, \bar{\nu}_\tau, \bar{\tau} \right\}$$

$$\mathbf{B_{97}} = \left\{ \nu_e, \bar{\nu}_e, u, d, c, \bar{s}, \nu_\tau, b \right\}$$

$$\mathbf{B}_{98} = \left\{ \nu_e, \bar{\nu}_e, u, \bar{u}, \nu_\mu, c, \tau, \bar{\tau} \right\}$$

$$\mathbf{B}_{99} = \left\{ \nu_e, \bar{\nu}_e, u, \bar{u}, \mu, \bar{\mu}, t, b \right\}$$

$$\mathbf{B}_{100} = \left\{ \nu_e, \bar{\nu}_e, u, \bar{u}, \bar{\nu}_\mu, \bar{s}, \bar{\nu}_\tau, \bar{t} \right\}$$

$$\mathbf{B}_{101} = \left\{ \nu_e, \bar{\nu}_e, u, \bar{u}, s, \bar{c}, \nu_\tau, \bar{b} \right\}$$

$$\mathbf{B}_{102} = \left\{ \nu_e, \bar{\nu}_e, u, \bar{d}, \nu_\mu, \mu, s, \bar{s} \right\}$$

$$\mathbf{B}_{103} = \left\{ \nu_e, \bar{\nu}_e, u, \bar{d}, \bar{\nu}_\mu, \bar{\mu}, c, \bar{c} \right\}$$

$$\mathbf{B}_{104} = \left\{ \nu_e, \bar{\nu}_e, u, \bar{d}, \nu_\tau, \tau, \bar{\nu}_\tau, t \right\}$$

$$\mathbf{B}_{105} = \left\{ \nu_e, \bar{\nu}_e, u, \bar{d}, \bar{\tau}, b, \bar{t}, \bar{b} \right\}$$

$$\mathbf{B}_{106} = \left\{ \nu_e, \bar{\nu}_e, d, \bar{u}, \nu_\mu, \bar{\mu}, \bar{c}, \bar{s} \right\}$$

$$\mathbf{B}_{107} = \left\{ \nu_e, \bar{\nu}_e, d, \bar{u}, \mu, \bar{\nu}_\mu, c, s \right\}$$

$$\mathbf{B}_{108} = \left\{ \nu_e, \bar{\nu}_e, d, \bar{u}, \nu_\tau, \bar{\tau}, t, \bar{t} \right\}$$

$$\mathbf{B}_{109} = \left\{ \nu_e, \bar{\nu}_e, d, \bar{u}, \tau, \bar{\nu}_\tau, b, \bar{b} \right\}$$

$$\mathbf{B}_{110} = \left\{ \nu_e, \bar{\nu}_e, d, \bar{d}, \nu_\mu, c, \bar{\nu}_\tau, \bar{t} \right\}$$

$$\mathbf{B}_{111} = \left\{ \nu_e, \bar{\nu}_e, d, \bar{d}, \mu, \bar{\mu}, \nu_\tau, \bar{b} \right\}$$

$$\mathbf{B}_{112} = \left\{ \nu_e, \bar{\nu}_e, d, \bar{d}, \bar{\nu}_\mu, \bar{s}, \tau, \bar{\tau} \right\}$$

$$\mathbf{B}_{113} = \left\{ \nu_e, \bar{\nu}_e, d, \bar{d}, s, \bar{c}, t, b \right\}$$

$$\mathbf{B}_{114} = \left\{ \nu_e, \bar{\nu}_e, \bar{u}, \bar{d}, \nu_\mu, \bar{\nu}_\mu, \nu_\tau, b \right\}$$

$$\mathbf{B}_{115} = \left\{ \nu_e, \bar{\nu}_e, \bar{u}, \bar{d}, \mu, \bar{c}, \bar{\nu}_\tau, \bar{\tau} \right\}$$

$$\mathbf{B}_{116} = \left\{ \nu_e, \bar{\nu}_e, \bar{u}, \bar{d}, \bar{\mu}, s, \tau, \bar{t} \right\}$$

$$\mathbf{B}_{117} = \left\{ \nu_e, \bar{\nu}_e, \bar{u}, \bar{d}, c, \bar{s}, t, \bar{b} \right\}$$

$$\mathbf{B}_{118} = \left\{ \nu_e, \bar{\nu}_e, \nu_\mu, \mu, \bar{\nu}_\mu, \bar{\mu}, \tau, \bar{\nu}_\tau \right\}$$

$$\mathbf{B}_{119} = \left\{ \nu_e, \bar{\nu}_e, \nu_\mu, \mu, c, \bar{c}, \nu_\tau, t \right\}$$

$$\mathbf{B}_{120} = \left\{ \nu_e, \bar{\nu}_e, \nu_\mu, \bar{\nu}_\mu, s, \bar{c}, \bar{\tau}, \bar{t} \right\}$$

$$\mathbf{B}_{121} = \left\{ \nu_e, \bar{\nu}_e, \nu_\mu, \bar{\mu}, c, s, b, \bar{b} \right\}$$

$$\mathbf{B}_{122} = \left\{ \nu_e, \bar{\nu}_e, \nu_\mu, \bar{s}, \nu_\tau, \bar{\nu}_\tau, \bar{\tau}, \bar{b} \right\}$$

$$\mathbf{B}_{123} = \left\{ \nu_e, \bar{\nu}_e, \nu_\mu, \bar{s}, \tau, t, b, \bar{t} \right\}$$

$$\mathbf{B}_{124} = \left\{ \nu_e, \bar{\nu}_e, \mu, \bar{\nu}_\mu, \bar{c}, \bar{s}, b, \bar{b} \right\}$$

$$\mathbf{B}_{125} = \left\{ \nu_e, \bar{\nu}_e, \mu, \bar{\mu}, c, \bar{s}, \bar{\tau}, \bar{t} \right\}$$

$$\mathbf{B}_{126} = \left\{ \nu_e, \bar{\nu}_e, \mu, s, \nu_\tau, \bar{\nu}_\tau, b, \bar{t} \right\}$$

$$\mathbf{B}_{127} = \left\{ \nu_e, \bar{\nu}_e, \mu, s, \tau, \bar{\tau}, t, \bar{b} \right\}$$

$$\mathbf{B}_{128} = \left\{ \nu_e, \bar{\nu}_e, \bar{\nu}_\mu, \bar{\mu}, s, \bar{s}, \nu_\tau, t \right\}$$

$$\mathbf{B}_{129} = \left\{ \nu_e, \bar{\nu}_e, \bar{\nu}_\mu, c, \nu_\tau, \tau, \bar{t}, \bar{b} \right\}$$

$$\mathbf{B}_{130} = \left\{ \nu_e, \bar{\nu}_e, \bar{\nu}_\mu, c, \bar{\nu}_\tau, \bar{\tau}, t, b \right\}$$

$$\mathbf{B}_{131} = \left\{ \nu_e, \bar{\nu}_e, \bar{\mu}, \bar{c}, \nu_\tau, \tau, \bar{\tau}, b \right\}$$

$$\mathbf{B}_{132} = \left\{ \nu_e, \bar{\nu}_e, \bar{\mu}, \bar{c}, \bar{\nu}_\tau, t, \bar{t}, \bar{b} \right\}$$

$$\mathbf{B}_{133} = \left\{ \nu_e, \bar{\nu}_e, c, s, \bar{c}, \bar{s}, \tau, \bar{\nu}_\tau \right\}$$

$$\mathbf{B}_{134} = \left\{ \nu_e, \bar{e}, u, d, \nu_\mu, \bar{s}, \tau, \bar{\nu}_\tau \right\}$$

$$\mathbf{B}_{135} = \left\{ \nu_e, \bar{e}, u, d, \mu, s, \nu_\tau, t \right\}$$

$$\mathbf{B}_{136} = \left\{ \nu_e, \bar{e}, u, d, \bar{\nu}_\mu, c, \bar{\tau}, \bar{t} \right\}$$

$$\mathbf{B}_{137} = \left\{ \nu_e, \bar{e}, u, d, \bar{\mu}, \bar{c}, b, \bar{b} \right\}$$

$$\mathbf{B}_{138} = \left\{ \nu_e, \bar{e}, u, \bar{u}, \nu_\mu, \mu, \bar{\nu}_\mu, \bar{c} \right\}$$

$$\mathbf{B}_{139} = \left\{ \nu_e, \bar{e}, u, \bar{u}, \bar{\mu}, c, s, \bar{s} \right\}$$

$$\mathbf{B}_{140} = \left\{ \nu_e, \bar{e}, u, \bar{u}, \nu_\tau, \bar{\nu}_\tau, \bar{\tau}, b \right\}$$

$$\mathbf{B}_{141} = \left\{ \nu_e, \bar{e}, u, \bar{u}, \tau, t, \bar{t}, \bar{b} \right\}$$

$$\mathbf{B}_{142} = \left\{ \nu_e, \bar{e}, u, \bar{d}, \nu_\mu, c, t, b \right\}$$

$$\mathbf{B}_{143} = \left\{ \nu_e, \bar{e}, u, \bar{d}, \mu, \bar{\mu}, \tau, \bar{\tau} \right\}$$

$$\mathbf{B}_{144} = \left\{ \nu_e, \bar{e}, u, \bar{d}, \bar{\nu}_\mu, \bar{s}, \nu_\tau, \bar{b} \right\}$$

$$\mathbf{B}_{145} = \left\{ \nu_e, \bar{e}, u, \bar{d}, s, \bar{c}, \bar{\nu}_\tau, \bar{t} \right\}$$

$$\mathbf{B}_{146} = \left\{ \nu_e, \bar{e}, d, \bar{u}, \nu_\mu, c, \nu_\tau, \bar{b} \right\}$$

$$\mathbf{B}_{147} = \left\{ \nu_e, \bar{e}, d, \bar{u}, \mu, \bar{\mu}, \bar{\nu}_\tau, \bar{t} \right\}$$

$$\mathbf{B}_{148} = \left\{ \nu_e, \bar{e}, d, \bar{u}, \bar{\nu}_\mu, \bar{s}, t, b \right\}$$

$$\mathbf{B}_{149} = \left\{ \nu_e, \bar{e}, d, \bar{u}, s, \bar{c}, \tau, \bar{\tau} \right\}$$

$$\mathbf{B}_{150} = \left\{ \nu_e, \bar{e}, d, \bar{d}, \nu_\mu, \bar{\nu}_\mu, \bar{\mu}, s \right\}$$

$$\mathbf{B}_{151} = \left\{ \nu_e, \bar{e}, d, \bar{d}, \mu, c, \bar{c}, \bar{s} \right\}$$

$$\mathbf{B}_{152} = \left\{ \nu_e, \bar{e}, d, \bar{d}, \nu_\tau, \tau, b, \bar{t} \right\}$$

$$\mathbf{B}_{153} = \left\{ \nu_e, \bar{e}, d, \bar{d}, \bar{\nu}_\tau, \bar{\tau}, t, \bar{b} \right\}$$

$$\mathbf{B}_{154} = \left\{ \nu_e, \bar{e}, \bar{u}, \bar{d}, \nu_\mu, \bar{s}, \bar{\tau}, \bar{t} \right\}$$

$$\mathbf{B}_{155} = \left\{ \nu_e, \bar{e}, \bar{u}, \bar{d}, \mu, s, b, \bar{b} \right\}$$

$$\mathbf{B}_{156} = \left\{ \nu_e, \bar{e}, \bar{u}, \bar{d}, \bar{\nu}_\mu, c, \tau, \bar{\nu}_\tau \right\}$$

$$\mathbf{B}_{157} = \left\{ \nu_e, \bar{e}, \bar{u}, \bar{d}, \bar{\mu}, \bar{c}, \nu_\tau, t \right\}$$

$$\mathbf{B}_{158} = \left\{ \nu_e, \bar{e}, \nu_\mu, \mu, \bar{\mu}, \bar{s}, t, \bar{b} \right\}$$

$$\mathbf{B}_{159} = \left\{ \nu_e, \bar{e}, \nu_\mu, \mu, c, s, \tau, \bar{t} \right\}$$

$$\mathbf{B}_{160} = \left\{ \nu_e, \bar{e}, \nu_\mu, \bar{\nu}_\mu, \nu_\tau, \tau, \bar{\tau}, t \right\}$$

$$\mathbf{B}_{161} = \left\{ \nu_e, \bar{e}, \nu_\mu, \bar{\nu}_\mu, \bar{\nu}_\tau, b, \bar{t}, \bar{b} \right\}$$

$$\mathbf{B}_{162} = \left\{ \nu_e, \bar{e}, \nu_\mu, \bar{\mu}, c, \bar{c}, \bar{\nu}_\tau, \bar{\tau} \right\}$$

$$\mathbf{B}_{163} = \left\{ \nu_e, \bar{e}, \nu_\mu, s, \bar{c}, \bar{s}, \nu_\tau, b \right\}$$

$$\mathbf{B}_{164} = \left\{ \nu_e, \bar{e}, \mu, \bar{\nu}_\mu, \bar{\mu}, c, \nu_\tau, b \right\}$$

$$\mathbf{B}_{165} = \left\{ \nu_e, \bar{e}, \mu, \bar{\nu}_\mu, s, \bar{s}, \bar{\nu}_\tau, \bar{\tau} \right\}$$

$$\mathbf{B}_{166} = \left\{ \nu_e, \bar{e}, \mu, \bar{c}, \nu_\tau, \bar{\tau}, \bar{t}, \bar{b} \right\}$$

$$\mathbf{B}_{167} = \left\{ \nu_e, \bar{e}, \mu, \bar{c}, \tau, \bar{\nu}_\tau, t, b \right\}$$

$$\mathbf{B}_{168} = \left\{ \nu_e, \bar{e}, \bar{\nu}_\mu, \bar{\mu}, \bar{c}, \bar{s}, \tau, \bar{t} \right\}$$

$$\mathbf{B}_{169} = \left\{ \nu_e, \bar{e}, \bar{\nu}_\mu, c, s, \bar{c}, t, \bar{b} \right\}$$

$$\mathbf{B}_{170} = \left\{ \nu_e, \bar{e}, \bar{\mu}, s, \nu_\tau, \tau, \bar{\nu}_\tau, \bar{b} \right\}$$

$$\mathbf{B}_{171} = \left\{ \nu_e, \bar{e}, \bar{\mu}, s, \bar{\tau}, t, b, \bar{t} \right\}$$

$$\mathbf{B}_{172} = \left\{ \nu_e, \bar{e}, c, \bar{s}, \nu_\tau, \bar{\nu}_\tau, t, \bar{t} \right\}$$

$$\mathbf{B}_{173} = \left\{ \nu_e, \bar{e}, c, \bar{s}, \tau, \bar{\tau}, b, \bar{b} \right\}$$

$$\mathbf{B}_{174} = \left\{ \nu_e, u, d, \bar{u}, \nu_\mu, s, b, \bar{t} \right\}$$

$$\mathbf{B}_{175} = \left\{ \nu_e, u, d, \bar{u}, \mu, \bar{s}, \bar{\tau}, \bar{b} \right\}$$

$$\mathbf{B}_{176} = \left\{ \nu_e, u, d, \bar{u}, \bar{\nu}_\mu, \bar{\mu}, \nu_\tau, \tau \right\}$$

$$\mathbf{B}_{177} = \left\{ \nu_e, u, d, \bar{u}, c, \bar{c}, \bar{\nu}_\tau, t \right\}$$

$$\mathbf{B}_{178} = \left\{ \nu_e, u, d, \bar{d}, \nu_\mu, \bar{c}, \nu_\tau, \bar{\tau} \right\}$$

$$\mathbf{B}_{179} = \left\{ \nu_e, u, d, \bar{d}, \mu, \bar{\nu}_\mu, \bar{\nu}_\tau, b \right\}$$

$$\mathbf{B}_{180} = \left\{ \nu_e, u, d, \bar{d}, \bar{\mu}, \bar{s}, t, \bar{t} \right\}$$

$$\mathbf{B}_{181} = \left\{ \nu_e, u, d, \bar{d}, c, s, \tau, \bar{b} \right\}$$

$$\mathbf{B}_{182} = \left\{ \nu_e, u, \bar{u}, \bar{d}, \nu_\mu, \bar{\mu}, \bar{\nu}_\tau, \bar{b} \right\}$$

$$\mathbf{B}_{183} = \left\{ \nu_e, u, \bar{u}, \bar{d}, \mu, c, \nu_\tau, \bar{t} \right\}$$

$$\mathbf{B}_{184} = \left\{ \nu_e, u, \bar{u}, \bar{d}, \bar{\nu}_\mu, s, \bar{\tau}, t \right\}$$

$$\mathbf{B}_{185} = \left\{ \nu_e, u, \bar{u}, \bar{d}, \bar{c}, \bar{s}, \tau, b \right\}$$

$$\mathbf{B}_{186} = \left\{ \nu_e, u, \nu_\mu, \mu, \nu_\tau, \tau, b, \bar{b} \right\}$$

$$\mathbf{B}_{187} = \left\{ \nu_e, u, \nu_\mu, \mu, \bar{\nu}_\tau, \bar{\tau}, t, \bar{t} \right\}$$

$$\mathbf{B}_{188} = \left\{ \nu_e, u, \nu_\mu, \bar{\nu}_\mu, \bar{\mu}, \bar{s}, \bar{\tau}, b \right\}$$

$$\mathbf{B}_{189} = \left\{ \nu_e, u, \nu_\mu, \bar{\nu}_\mu, c, s, \nu_\tau, \bar{\nu}_\tau \right\}$$

$$\mathbf{B}_{190} = \left\{ \nu_e, u, \nu_\mu, \bar{\mu}, s, \bar{c}, \tau, t \right\}$$

$$\mathbf{B}_{191} = \left\{ \nu_e, u, \nu_\mu, c, \bar{c}, \bar{s}, \bar{t}, \bar{b} \right\}$$

$$\mathbf{B}_{192} = \left\{ \nu_e, u, \mu, \bar{\nu}_\mu, \bar{\mu}, s, \bar{t}, \bar{b} \right\}$$

$$\mathbf{B}_{193} = \left\{ \nu_e, u, \mu, \bar{\nu}_\mu, c, \bar{s}, \tau, t \right\}$$

$$\mathbf{B}_{194} = \left\{ \nu_e, u, \mu, \bar{\mu}, \bar{c}, \bar{s}, \nu_\tau, \bar{\nu}_\tau \right\}$$

$$\mathbf{B}_{195} = \left\{ \nu_e, u, \mu, c, s, \bar{c}, \bar{\tau}, b \right\}$$

$$\mathbf{B}_{196} = \left\{ \nu_e, u, \bar{\nu}_\mu, \bar{c}, \nu_\tau, t, b, \bar{t} \right\}$$

$$\mathbf{B}_{197} = \left\{ \nu_e, u, \bar{\nu}_\mu, \bar{c}, \tau, \bar{\nu}_\tau, \bar{\tau}, \bar{b} \right\}$$

$$\mathbf{B}_{198} = \left\{ \nu_e, u, \bar{\mu}, c, \nu_\tau, \bar{\tau}, t, \bar{b} \right\}$$

$$\mathbf{B}_{199} = \left\{ \nu_e, u, \bar{\mu}, c, \tau, \bar{\nu}_\tau, b, \bar{t} \right\}$$

$$\mathbf{B}_{200} = \left\{ \nu_e, u, s, \bar{s}, \nu_\tau, \tau, \bar{\tau}, \bar{t} \right\}$$

$$\mathbf{B}_{201} = \left\{ \nu_e, u, s, \bar{s}, \bar{\nu}_\tau, t, b, \bar{b} \right\}$$

$$\mathbf{B}_{202} = \left\{ \nu_e, d, \bar{u}, \bar{d}, \nu_\mu, \mu, \tau, t \right\}$$

$$\mathbf{B}_{203} = \left\{ \nu_e, d, \bar{u}, \bar{d}, \bar{\nu}_\mu, \bar{c}, \bar{t}, \bar{b} \right\}$$

$$\mathbf{B}_{204} = \left\{ \nu_e, d, \bar{u}, \bar{d}, \bar{\mu}, c, \bar{\tau}, b \right\}$$

$$\mathbf{B}_{205} = \left\{ \nu_e, d, \bar{u}, \bar{d}, s, \bar{s}, \nu_\tau, \bar{\nu}_\tau \right\}$$

$$\mathbf{B}_{206} = \left\{ \nu_e, d, \nu_\mu, \mu, \bar{\nu}_\mu, \bar{s}, \nu_\tau, \bar{t} \right\}$$

$$\mathbf{B}_{207} = \left\{ \nu_e, d, \nu_\mu, \mu, s, \bar{c}, \bar{\nu}_\tau, \bar{b} \right\}$$

$$\mathbf{B}_{208} = \left\{ \nu_e, d, \nu_\mu, \bar{\nu}_\mu, c, \bar{c}, \tau, b \right\}$$

$$\mathbf{B}_{209} = \left\{ \nu_e, d, \nu_\mu, \bar{\mu}, \nu_\tau, \bar{\nu}_\tau, t, b \right\}$$

$$\mathbf{B}_{210} = \left\{ \nu_e, d, \nu_\mu, \bar{\mu}, \tau, \bar{\tau}, \bar{t}, \bar{b} \right\}$$

$$\mathbf{B}_{211} = \left\{ \nu_e, d, \nu_\mu, c, s, \bar{s}, \bar{\tau}, t \right\}$$

$$\mathbf{B}_{212} = \left\{ \nu_e, d, \mu, \bar{\nu}_\mu, \bar{\mu}, \bar{c}, \bar{\tau}, t \right\}$$

$$\mathbf{B}_{213} = \left\{ \nu_e, d, \mu, \bar{\mu}, s, \bar{s}, \tau, b \right\}$$

$$\mathbf{B}_{214} = \left\{ \nu_e, d, \mu, c, \nu_\tau, \tau, \bar{\nu}_\tau, \bar{\tau} \right\}$$

$$\mathbf{B}_{215} = \left\{ \nu_e, d, \mu, c, t, b, \bar{t}, \bar{b} \right\}$$

$$\mathbf{B}_{216} = \left\{ \nu_e, d, \bar{\nu}_\mu, \bar{\mu}, c, \bar{s}, \bar{\nu}_\tau, \bar{b} \right\}$$

$$\mathbf{B}_{217} = \left\{ \nu_e, d, \bar{\nu}_\mu, s, \nu_\tau, \bar{\tau}, b, \bar{b} \right\}$$

$$\mathbf{B}_{218} = \left\{ \nu_e, d, \bar{\nu}_\mu, s, \tau, \bar{\nu}_\tau, t, \bar{t} \right\}$$

$$\mathbf{B}_{219} = \left\{ \nu_e, d, \bar{\mu}, c, s, \bar{c}, \nu_\tau, \bar{t} \right\}$$

$$\mathbf{B}_{220} = \left\{ \nu_e, d, \bar{c}, \bar{s}, \nu_\tau, \tau, t, \bar{b} \right\}$$

$$\mathbf{B}_{221} = \left\{ \nu_e, d, \bar{c}, \bar{s}, \bar{\nu}_\tau, \bar{\tau}, b, \bar{t} \right\}$$

$$\mathbf{B}_{222} = \left\{ \nu_e, \bar{u}, \nu_\mu, \mu, \bar{\mu}, s, \nu_\tau, \bar{\tau} \right\}$$

$$\mathbf{B}_{223} = \left\{ \nu_e, \bar{u}, \nu_\mu, \mu, c, \bar{s}, \bar{\nu}_\tau, b \right\}$$

$$\mathbf{B}_{224} = \left\{ \nu_e, \bar{u}, \nu_\mu, \bar{\nu}_\mu, \bar{\mu}, c, t, \bar{t} \right\}$$

$$\mathbf{B}_{225} = \left\{ \nu_e, \bar{u}, \nu_\mu, \bar{\nu}_\mu, s, \bar{s}, \tau, \bar{b} \right\}$$

$$\mathbf{B}_{226} = \left\{ \nu_e, \bar{u}, \nu_\mu, \bar{c}, \nu_\tau, \tau, \bar{\nu}_\tau, \bar{t} \right\}$$

$$\mathbf{B}_{227} = \left\{ \nu_e, \bar{u}, \nu_\mu, \bar{c}, \bar{\tau}, t, b, \bar{b} \right\}$$

$$\mathbf{B}_{228} = \left\{ \nu_e, \bar{u}, \mu, \bar{\nu}_\mu, \nu_\tau, \bar{\nu}_\tau, t, \bar{b} \right\}$$

$$\mathbf{B}_{229} = \left\{ \nu_e, \bar{u}, \mu, \bar{\nu}_\mu, \tau, \bar{\tau}, b, \bar{t} \right\}$$

$$\mathbf{B}_{230} = \left\{ \nu_e, \bar{u}, \mu, \bar{\mu}, c, \bar{c}, \tau, \bar{b} \right\}$$

$$\mathbf{B}_{231} = \left\{ \nu_e, \bar{u}, \mu, s, \bar{c}, \bar{s}, t, \bar{t} \right\}$$

$$\mathbf{B}_{232} = \left\{ \nu_e, \bar{u}, \bar{\nu}_\mu, \bar{\mu}, s, \bar{c}, \bar{\nu}_\tau, b \right\}$$

$$\mathbf{B}_{233} = \left\{ \nu_e, \bar{u}, \bar{\nu}_\mu, c, \bar{c}, \bar{s}, \nu_\tau, \bar{\tau} \right\}$$

$$\mathbf{B}_{234} = \left\{ \nu_e, \bar{u}, \bar{\mu}, \bar{s}, \nu_\tau, b, \bar{t}, \bar{b} \right\}$$

$$\mathbf{B}_{235} = \left\{ \nu_e, \bar{u}, \bar{\mu}, \bar{s}, \tau, \bar{\nu}_\tau, \bar{\tau}, t \right\}$$

$$\mathbf{B}_{236} = \left\{ \nu_e, \bar{u}, c, s, \nu_\tau, \tau, t, b \right\}$$

$$\mathbf{B}_{237} = \left\{ \nu_e, \bar{u}, c, s, \bar{\nu}_\tau, \bar{\tau}, \bar{t}, \bar{b} \right\}$$

$$\mathbf{B_{238}} = \left\{ \nu_e , \bar{d} , \nu_\mu , \mu , \bar{\nu}_\mu , c , \bar{\tau} , \bar{b} \right\}$$

$$\mathbf{B_{239}} = \left\{ \nu_e , \bar{d} , \nu_\mu , \mu , \bar{\mu} , \bar{c} , b , \bar{t} \right\}$$

$$\mathbf{B_{240}} = \left\{ \nu_e , \bar{d} , \nu_\mu , \bar{\nu}_\mu , \bar{c} , \bar{s} , \bar{\nu}_\tau , t \right\}$$

$$\mathbf{B_{241}} = \left\{ \nu_e , \bar{d} , \nu_\mu , \bar{\mu} , c , \bar{s} , \nu_\tau , \tau \right\}$$

$$\mathbf{B_{242}} = \left\{ \nu_e , \bar{d} , \nu_\mu , s , \nu_\tau , t , \bar{t} , \bar{b} \right\}$$

$$\mathbf{B_{243}} = \left\{ \nu_e , \bar{d} , \nu_\mu , s , \tau , \bar{\nu}_\tau , \bar{\tau} , b \right\}$$

$$\mathbf{B_{244}} = \left\{ \nu_e , \bar{d} , \mu , \bar{\nu}_\mu , s , \bar{c} , \nu_\tau , \tau \right\}$$

$$\mathbf{B_{245}} = \left\{ \nu_e , \bar{d} , \mu , \bar{\mu} , c , s , \bar{\nu}_\tau , t \right\}$$

$$\mathbf{B_{246}} = \left\{ \nu_e , \bar{d} , \mu , \bar{s} , \nu_\tau , \bar{\tau} , t , b \right\}$$

$$\mathbf{B_{247}} = \left\{ \nu_e , \bar{d} , \mu , \bar{s} , \tau , \bar{\nu}_\tau , \bar{t} , \bar{b} \right\}$$

$$\mathbf{B_{248}} = \left\{ \nu_e , \bar{d} , \bar{\nu}_\mu , \bar{\mu} , \nu_\tau , \bar{\nu}_\tau , \bar{\tau} , \bar{t} \right\}$$

$$\mathbf{B_{249}} = \left\{ \nu_e , \bar{d} , \bar{\nu}_\mu , \bar{\mu} , \tau , t , b , \bar{b} \right\}$$

$$\mathbf{B_{250}} = \left\{ \nu_e , \bar{d} , \bar{\nu}_\mu , c , s , \bar{s} , b , \bar{t} \right\}$$

$$\mathbf{B_{251}} = \left\{ \nu_e , \bar{d} , \bar{\mu} , s , \bar{c} , \bar{s} , \bar{\tau} , \bar{b} \right\}$$

$$\mathbf{B}_{252} = \left\{ \nu_e, \bar{d}, c, \bar{c}, \nu_\tau, \bar{\nu}_\tau, b, \bar{b} \right\}$$

$$\mathbf{B}_{253} = \left\{ \nu_e, \bar{d}, c, \bar{c}, \tau, \bar{\tau}, t, \bar{t} \right\}$$

$$\mathbf{B}_{254} = \left\{ e, \bar{\nu}_e, \bar{e}, u, \nu_\mu, \mu, \tau, t \right\}$$

$$\mathbf{B}_{255} = \left\{ e, \bar{\nu}_e, \bar{e}, u, \bar{\nu}_\mu, \bar{c}, \bar{t}, \bar{b} \right\}$$

$$\mathbf{B}_{256} = \left\{ e, \bar{\nu}_e, \bar{e}, u, \bar{\mu}, c, \bar{\tau}, b \right\}$$

$$\mathbf{B}_{257} = \left\{ e, \bar{\nu}_e, \bar{e}, u, s, \bar{s}, \nu_\tau, \bar{\nu}_\tau \right\}$$

$$\mathbf{B}_{258} = \left\{ e, \bar{\nu}_e, \bar{e}, d, \nu_\mu, \bar{\mu}, \bar{\nu}_\tau, \bar{b} \right\}$$

$$\mathbf{B}_{259} = \left\{ e, \bar{\nu}_e, \bar{e}, d, \mu, c, \nu_\tau, \bar{t} \right\}$$

$$\mathbf{B}_{260} = \left\{ e, \bar{\nu}_e, \bar{e}, d, \bar{\nu}_\mu, s, \bar{\tau}, t \right\}$$

$$\mathbf{B}_{261} = \left\{ e, \bar{\nu}_e, \bar{e}, d, \bar{c}, \bar{s}, \tau, b \right\}$$

$$\mathbf{B}_{262} = \left\{ e, \bar{\nu}_e, \bar{e}, \bar{u}, \nu_\mu, \bar{c}, \nu_\tau, \bar{\tau} \right\}$$

$$\mathbf{B}_{263} = \left\{ e, \bar{\nu}_e, \bar{e}, \bar{u}, \mu, \bar{\nu}_\mu, \bar{\nu}_\tau, b \right\}$$

$$\mathbf{B}_{264} = \left\{ e, \bar{\nu}_e, \bar{e}, \bar{u}, \bar{\mu}, \bar{s}, t, \bar{t} \right\}$$

$$\mathbf{B}_{265} = \left\{ e, \bar{\nu}_e, \bar{e}, \bar{u}, c, s, \tau, \bar{b} \right\}$$

$$\mathbf{B}_{266} = \left\{ e, \bar{\nu}_e, \bar{e}, \bar{d}, \nu_\mu, s, b, \bar{t} \right\}$$

$$\mathbf{B}_{267} = \left\{ e, \bar{\nu}_e, \bar{e}, \bar{d}, \mu, \bar{s}, \bar{\tau}, \bar{b} \right\}$$

$$\mathbf{B}_{268} = \left\{ e, \bar{\nu}_e, \bar{e}, \bar{d}, \bar{\nu}_\mu, \bar{\mu}, \nu_\tau, \tau \right\}$$

$$\mathbf{B}_{269} = \left\{ e, \bar{\nu}_e, \bar{e}, \bar{d}, c, \bar{c}, \bar{\nu}_\tau, t \right\}$$

$$\mathbf{B}_{270} = \left\{ e, \bar{\nu}_e, u, d, \nu_\mu, \bar{s}, \bar{\tau}, \bar{t} \right\}$$

$$\mathbf{B}_{271} = \left\{ e, \bar{\nu}_e, u, d, \mu, s, b, \bar{b} \right\}$$

$$\mathbf{B}_{272} = \left\{ e, \bar{\nu}_e, u, d, \bar{\nu}_\mu, c, \tau, \bar{\nu}_\tau \right\}$$

$$\mathbf{B}_{273} = \left\{ e, \bar{\nu}_e, u, d, \bar{\mu}, \bar{c}, \nu_\tau, t \right\}$$

$$\mathbf{B}_{274} = \left\{ e, \bar{\nu}_e, u, \bar{u}, \nu_\mu, \bar{\nu}_\mu, \bar{\mu}, s \right\}$$

$$\mathbf{B}_{275} = \left\{ e, \bar{\nu}_e, u, \bar{u}, \mu, c, \bar{c}, \bar{s} \right\}$$

$$\mathbf{B}_{276} = \left\{ e, \bar{\nu}_e, u, \bar{u}, \nu_\tau, \tau, b, \bar{t} \right\}$$

$$\mathbf{B}_{277} = \left\{ e, \bar{\nu}_e, u, \bar{u}, \bar{\nu}_\tau, \bar{\tau}, t, \bar{b} \right\}$$

$$\mathbf{B}_{278} = \left\{ e, \bar{\nu}_e, u, \bar{d}, \nu_\mu, c, \nu_\tau, \bar{b} \right\}$$

$$\mathbf{B}_{279} = \left\{ e, \bar{\nu}_e, u, \bar{d}, \mu, \bar{\mu}, \bar{\nu}_\tau, \bar{t} \right\}$$

$$\mathbf{B}_{280} = \left\{ e, \bar{\nu}_e, u, \bar{d}, \bar{\nu}_\mu, \bar{s}, t, b \right\}$$

$$\mathbf{B}_{281} = \left\{ e, \bar{\nu}_e, u, \bar{d}, s, \bar{c}, \tau, \bar{\tau} \right\}$$

$$\mathbf{B}_{282} = \left\{ e, \bar{\nu}_e, d, \bar{u}, \nu_\mu, c, t, b \right\}$$

$$\mathbf{B}_{283} = \left\{ e, \bar{\nu}_e, d, \bar{u}, \mu, \bar{\mu}, \tau, \bar{\tau} \right\}$$

$$\mathbf{B}_{284} = \left\{ e, \bar{\nu}_e, d, \bar{u}, \bar{\nu}_\mu, \bar{s}, \nu_\tau, \bar{b} \right\}$$

$$\mathbf{B}_{285} = \left\{ e, \bar{\nu}_e, d, \bar{u}, s, \bar{c}, \bar{\nu}_\tau, \bar{t} \right\}$$

$$\mathbf{B}_{286} = \left\{ e, \bar{\nu}_e, d, \bar{d}, \nu_\mu, \mu, \bar{\nu}_\mu, \bar{c} \right\}$$

$$\mathbf{B}_{287} = \left\{ e, \bar{\nu}_e, d, \bar{d}, \bar{\mu}, c, s, \bar{s} \right\}$$

$$\mathbf{B}_{288} = \left\{ e, \bar{\nu}_e, d, \bar{d}, \nu_\tau, \bar{\nu}_\tau, \bar{\tau}, b \right\}$$

$$\mathbf{B}_{289} = \left\{ e, \bar{\nu}_e, d, \bar{d}, \tau, t, \bar{t}, \bar{b} \right\}$$

$$\mathbf{B}_{290} = \left\{ e, \bar{\nu}_e, \bar{u}, \bar{d}, \nu_\mu, \bar{s}, \tau, \bar{\nu}_\tau \right\}$$

$$\mathbf{B}_{291} = \left\{ e, \bar{\nu}_e, \bar{u}, \bar{d}, \mu, s, \nu_\tau, t \right\}$$

$$\mathbf{B}_{292} = \left\{ e, \bar{\nu}_e, \bar{u}, \bar{d}, \bar{\nu}_\mu, c, \bar{\tau}, \bar{t} \right\}$$

$$\mathbf{B}_{293} = \left\{ e, \bar{\nu}_e, \bar{u}, \bar{d}, \bar{\mu}, \bar{c}, b, \bar{b} \right\}$$

$$\mathbf{B_{294}} = \left\{ e, \bar{\nu}_e, \nu_\mu, \mu, \bar{\mu}, \bar{s}, \nu_\tau, b \right\}$$

$$\mathbf{B_{295}} = \left\{ e, \bar{\nu}_e, \nu_\mu, \mu, c, s, \bar{\nu}_\tau, \bar{\tau} \right\}$$

$$\mathbf{B_{296}} = \left\{ e, \bar{\nu}_e, \nu_\mu, \bar{\nu}_\mu, \nu_\tau, \bar{\nu}_\tau, t, \bar{t} \right\}$$

$$\mathbf{B_{297}} = \left\{ e, \bar{\nu}_e, \nu_\mu, \bar{\nu}_\mu, \tau, \bar{\tau}, b, \bar{b} \right\}$$

$$\mathbf{B_{298}} = \left\{ e, \bar{\nu}_e, \nu_\mu, \bar{\mu}, c, \bar{c}, \tau, \bar{t} \right\}$$

$$\mathbf{B_{299}} = \left\{ e, \bar{\nu}_e, \nu_\mu, s, \bar{c}, \bar{s}, t, \bar{b} \right\}$$

$$\mathbf{B_{300}} = \left\{ e, \bar{\nu}_e, \mu, \bar{\nu}_\mu, \bar{\mu}, c, t, \bar{b} \right\}$$

$$\mathbf{B_{301}} = \left\{ e, \bar{\nu}_e, \mu, \bar{\nu}_\mu, s, \bar{s}, \tau, \bar{t} \right\}$$

$$\mathbf{B_{302}} = \left\{ e, \bar{\nu}_e, \mu, \bar{c}, \nu_\tau, \tau, \bar{\nu}_\tau, \bar{b} \right\}$$

$$\mathbf{B_{303}} = \left\{ e, \bar{\nu}_e, \mu, \bar{c}, \bar{\tau}, t, b, \bar{t} \right\}$$

$$\mathbf{B_{304}} = \left\{ e, \bar{\nu}_e, \bar{\nu}_\mu, \bar{\mu}, \bar{c}, \bar{s}, \bar{\nu}_\tau, \bar{\tau} \right\}$$

$$\mathbf{B_{305}} = \left\{ e, \bar{\nu}_e, \bar{\nu}_\mu, c, s, \bar{c}, \nu_\tau, b \right\}$$

$$\mathbf{B_{306}} = \left\{ e, \bar{\nu}_e, \bar{\mu}, s, \nu_\tau, \bar{\tau}, \bar{t}, \bar{b} \right\}$$

$$\mathbf{B_{307}} = \left\{ e, \bar{\nu}_e, \bar{\mu}, s, \tau, \bar{\nu}_\tau, t, b \right\}$$

$$\mathbf{B}_{308} = \left\{ e, \bar{\nu}_e, c, \bar{s}, \nu_\tau, \tau, \bar{\tau}, t \right\}$$

$$\mathbf{B}_{309} = \left\{ e, \bar{\nu}_e, c, \bar{s}, \bar{\nu}_\tau, b, \bar{t}, \bar{b} \right\}$$

$$\mathbf{B}_{310} = \left\{ e, \bar{e}, u, d, \nu_\mu, \bar{\nu}_\mu, \nu_\tau, b \right\}$$

$$\mathbf{B}_{311} = \left\{ e, \bar{e}, u, d, \mu, \bar{c}, \bar{\nu}_\tau, \bar{\tau} \right\}$$

$$\mathbf{B}_{312} = \left\{ e, \bar{e}, u, d, \bar{\mu}, s, \tau, \bar{t} \right\}$$

$$\mathbf{B}_{313} = \left\{ e, \bar{e}, u, d, c, \bar{s}, t, \bar{b} \right\}$$

$$\mathbf{B}_{314} = \left\{ e, \bar{e}, u, \bar{u}, \nu_\mu, c, \bar{\nu}_\tau, \bar{t} \right\}$$

$$\mathbf{B}_{315} = \left\{ e, \bar{e}, u, \bar{u}, \mu, \bar{\mu}, \nu_\tau, \bar{b} \right\}$$

$$\mathbf{B}_{316} = \left\{ e, \bar{e}, u, \bar{u}, \bar{\nu}_\mu, \bar{s}, \tau, \bar{\tau} \right\}$$

$$\mathbf{B}_{317} = \left\{ e, \bar{e}, u, \bar{u}, s, \bar{c}, t, b \right\}$$

$$\mathbf{B}_{318} = \left\{ e, \bar{e}, u, \bar{d}, \nu_\mu, \bar{\mu}, \bar{c}, \bar{s} \right\}$$

$$\mathbf{B}_{319} = \left\{ e, \bar{e}, u, \bar{d}, \mu, \bar{\nu}_\mu, c, s \right\}$$

$$\mathbf{B}_{320} = \left\{ e, \bar{e}, u, \bar{d}, \nu_\tau, \bar{\tau}, t, \bar{t} \right\}$$

$$\mathbf{B}_{321} = \left\{ e, \bar{e}, u, \bar{d}, \tau, \bar{\nu}_\tau, b, \bar{b} \right\}$$

$$\mathbf{B}_{322} = \left\{ e, \bar{e}, d, \bar{u}, \nu_\mu, \mu, s, \bar{s} \right\}$$

$$\mathbf{B}_{323} = \left\{ e, \bar{e}, d, \bar{u}, \bar{\nu}_\mu, \bar{\mu}, c, \bar{c} \right\}$$

$$\mathbf{B}_{324} = \left\{ e, \bar{e}, d, \bar{u}, \nu_\tau, \tau, \bar{\nu}_\tau, t \right\}$$

$$\mathbf{B}_{325} = \left\{ e, \bar{e}, d, \bar{u}, \bar{\tau}, b, \bar{t}, \bar{b} \right\}$$

$$\mathbf{B}_{326} = \left\{ e, \bar{e}, d, \bar{d}, \nu_\mu, c, \tau, \bar{\tau} \right\}$$

$$\mathbf{B}_{327} = \left\{ e, \bar{e}, d, \bar{d}, \mu, \bar{\mu}, t, b \right\}$$

$$\mathbf{B}_{328} = \left\{ e, \bar{e}, d, \bar{d}, \bar{\nu}_\mu, \bar{s}, \bar{\nu}_\tau, \bar{t} \right\}$$

$$\mathbf{B}_{329} = \left\{ e, \bar{e}, d, \bar{d}, s, \bar{c}, \nu_\tau, \bar{b} \right\}$$

$$\mathbf{B}_{330} = \left\{ e, \bar{e}, \bar{u}, \bar{d}, \nu_\mu, \bar{\nu}_\mu, t, \bar{b} \right\}$$

$$\mathbf{B}_{331} = \left\{ e, \bar{e}, \bar{u}, \bar{d}, \mu, \bar{c}, \tau, \bar{t} \right\}$$

$$\mathbf{B}_{332} = \left\{ e, \bar{e}, \bar{u}, \bar{d}, \bar{\mu}, s, \bar{\nu}_\tau, \bar{\tau} \right\}$$

$$\mathbf{B}_{333} = \left\{ e, \bar{e}, \bar{u}, \bar{d}, c, \bar{s}, \nu_\tau, b \right\}$$

$$\mathbf{B}_{334} = \left\{ e, \bar{e}, \nu_\mu, \mu, \bar{\nu}_\mu, \bar{\mu}, \bar{\tau}, \bar{t} \right\}$$

$$\mathbf{B}_{335} = \left\{ e, \bar{e}, \nu_\mu, \mu, c, \bar{c}, b, \bar{b} \right\}$$

$$\mathbf{B}_{336} = \left\{ e, \bar{e}, \nu_\mu, \bar{\nu}_\mu, s, \bar{c}, \tau, \bar{\nu}_\tau \right\}$$

$$\mathbf{B}_{337} = \left\{ e, \bar{e}, \nu_\mu, \bar{\mu}, c, s, \nu_\tau, t \right\}$$

$$\mathbf{B}_{338} = \left\{ e, \bar{e}, \nu_\mu, \bar{s}, \nu_\tau, \tau, \bar{t}, \bar{b} \right\}$$

$$\mathbf{B}_{339} = \left\{ e, \bar{e}, \nu_\mu, \bar{s}, \bar{\nu}_\tau, \bar{\tau}, t, b \right\}$$

$$\mathbf{B}_{340} = \left\{ e, \bar{e}, \mu, \bar{\nu}_\mu, \bar{c}, \bar{s}, \nu_\tau, t \right\}$$

$$\mathbf{B}_{341} = \left\{ e, \bar{e}, \mu, \bar{\mu}, c, \bar{s}, \tau, \bar{\nu}_\tau \right\}$$

$$\mathbf{B}_{342} = \left\{ e, \bar{e}, \mu, s, \nu_\tau, \tau, \bar{\tau}, b \right\}$$

$$\mathbf{B}_{343} = \left\{ e, \bar{e}, \mu, s, \bar{\nu}_\tau, t, \bar{t}, \bar{b} \right\}$$

$$\mathbf{B}_{344} = \left\{ e, \bar{e}, \bar{\nu}_\mu, \bar{\mu}, s, \bar{s}, b, \bar{b} \right\}$$

$$\mathbf{B}_{345} = \left\{ e, \bar{e}, \bar{\nu}_\mu, c, \nu_\tau, \bar{\nu}_\tau, \bar{\tau}, \bar{b} \right\}$$

$$\mathbf{B}_{346} = \left\{ e, \bar{e}, \bar{\nu}_\mu, c, \tau, t, b, \bar{t} \right\}$$

$$\mathbf{B}_{347} = \left\{ e, \bar{e}, \bar{\mu}, \bar{c}, \nu_\tau, \bar{\nu}_\tau, b, \bar{t} \right\}$$

$$\mathbf{B}_{348} = \left\{ e, \bar{e}, \bar{\mu}, \bar{c}, \tau, \bar{\tau}, t, \bar{b} \right\}$$

$$\mathbf{B}_{349} = \left\{ e, \bar{e}, c, s, \bar{c}, \bar{s}, \bar{\tau}, \bar{t} \right\}$$

$$\mathbf{B}_{350} = \left\{ e, u, d, \bar{u}, \nu_\mu, \bar{c}, \tau, \bar{b} \right\}$$

$$\mathbf{B}_{351} = \left\{ e, u, d, \bar{u}, \mu, \bar{\nu}_\mu, t, \bar{t} \right\}$$

$$\mathbf{B}_{352} = \left\{ e, u, d, \bar{u}, \bar{\mu}, \bar{s}, \bar{\nu}_\tau, b \right\}$$

$$\mathbf{B}_{353} = \left\{ e, u, d, \bar{u}, c, s, \nu_\tau, \bar{\tau} \right\}$$

$$\mathbf{B}_{354} = \left\{ e, u, d, \bar{d}, \nu_\mu, s, \bar{\nu}_\tau, t \right\}$$

$$\mathbf{B}_{355} = \left\{ e, u, d, \bar{d}, \mu, \bar{s}, \nu_\tau, \tau \right\}$$

$$\mathbf{B}_{356} = \left\{ e, u, d, \bar{d}, \bar{\nu}_\mu, \bar{\mu}, \bar{\tau}, \bar{b} \right\}$$

$$\mathbf{B}_{357} = \left\{ e, u, d, \bar{d}, c, \bar{c}, b, \bar{t} \right\}$$

$$\mathbf{B}_{358} = \left\{ e, u, \bar{u}, \bar{d}, \nu_\mu, \mu, \bar{\tau}, b \right\}$$

$$\mathbf{B}_{359} = \left\{ e, u, \bar{u}, \bar{d}, \bar{\nu}_\mu, \bar{c}, \nu_\tau, \bar{\nu}_\tau \right\}$$

$$\mathbf{B}_{360} = \left\{ e, u, \bar{u}, \bar{d}, \bar{\mu}, c, \tau, t \right\}$$

$$\mathbf{B}_{361} = \left\{ e, u, \bar{u}, \bar{d}, s, \bar{s}, \bar{t}, \bar{b} \right\}$$

$$\mathbf{B}_{362} = \left\{ e, u, \nu_\mu, \mu, \bar{\nu}_\mu, \bar{s}, \bar{\nu}_\tau, \bar{b} \right\}$$

$$\mathbf{B}_{363} = \left\{ e, u, \nu_\mu, \mu, s, \bar{c}, \nu_\tau, \bar{t} \right\}$$

$$\mathbf{B}_{364} = \left\{ e, u, \nu_\mu, \bar{\nu}_\mu, c, \bar{c}, \bar{\tau}, t \right\}$$

$$\mathbf{B}_{365} = \left\{ e, u, \nu_\mu, \bar{\mu}, \nu_\tau, \tau, \bar{\nu}_\tau, \bar{\tau} \right\}$$

$$\mathbf{B}_{366} = \left\{ e, u, \nu_\mu, \bar{\mu}, t, b, \bar{t}, \bar{b} \right\}$$

$$\mathbf{B}_{367} = \left\{ e, u, \nu_\mu, c, s, \bar{s}, \tau, b \right\}$$

$$\mathbf{B}_{368} = \left\{ e, u, \mu, \bar{\nu}_\mu, \bar{\mu}, \bar{c}, \tau, b \right\}$$

$$\mathbf{B}_{369} = \left\{ e, u, \mu, \bar{\mu}, s, \bar{s}, \bar{\tau}, t \right\}$$

$$\mathbf{B}_{370} = \left\{ e, u, \mu, c, \nu_\tau, \bar{\nu}_\tau, t, b \right\}$$

$$\mathbf{B}_{371} = \left\{ e, u, \mu, c, \tau, \bar{\tau}, \bar{t}, \bar{b} \right\}$$

$$\mathbf{B}_{372} = \left\{ e, u, \bar{\nu}_\mu, \bar{\mu}, c, \bar{s}, \nu_\tau, \bar{t} \right\}$$

$$\mathbf{B}_{373} = \left\{ e, u, \bar{\nu}_\mu, s, \nu_\tau, \tau, t, \bar{b} \right\}$$

$$\mathbf{B}_{374} = \left\{ e, u, \bar{\nu}_\mu, s, \bar{\nu}_\tau, \bar{\tau}, b, \bar{t} \right\}$$

$$\mathbf{B}_{375} = \left\{ e, u, \bar{\mu}, c, s, \bar{c}, \bar{\nu}_\tau, \bar{b} \right\}$$

$$\mathbf{B}_{376} = \left\{ e, u, \bar{c}, \bar{s}, \nu_\tau, \bar{\tau}, b, \bar{b} \right\}$$

$$\mathbf{B}_{377} = \left\{ e, u, \bar{c}, \bar{s}, \tau, \bar{\nu}_\tau, t, \bar{t} \right\}$$

$$\mathbf{B}_{378} = \left\{ e \, , d \, , \bar{u} \, , \bar{d} \, , \nu_\mu \, , \bar{\mu} \, , \nu_\tau \, , \bar{t} \right\}$$

$$\mathbf{B}_{379} = \left\{ e \, , d \, , \bar{u} \, , \bar{d} \, , \mu \, , c \, , \bar{\nu}_\tau \, , \bar{b} \right\}$$

$$\mathbf{B}_{380} = \left\{ e \, , d \, , \bar{u} \, , \bar{d} \, , \bar{\nu}_\mu \, , s \, , \tau \, , b \right\}$$

$$\mathbf{B}_{381} = \left\{ e \, , d \, , \bar{u} \, , \bar{d} \, , \bar{c} \, , \bar{s} \, , \bar{\tau} \, , t \right\}$$

$$\mathbf{B}_{382} = \left\{ e \, , d \, , \nu_\mu \, , \mu \, , \nu_\tau \, , \bar{\tau} \, , t \, , \bar{b} \right\}$$

$$\mathbf{B}_{383} = \left\{ e \, , d \, , \nu_\mu \, , \mu \, , \tau \, , \bar{\nu}_\tau \, , b \, , \bar{t} \right\}$$

$$\mathbf{B}_{384} = \left\{ e \, , d \, , \nu_\mu \, , \bar{\nu}_\mu \, , \bar{\mu} \, , \bar{s} \, , \tau \, , t \right\}$$

$$\mathbf{B}_{385} = \left\{ e \, , d \, , \nu_\mu \, , \bar{\nu}_\mu \, , c \, , s \, , \bar{t} \, , \bar{b} \right\}$$

$$\mathbf{B}_{386} = \left\{ e \, , d \, , \nu_\mu \, , \bar{\mu} \, , s \, , \bar{c} \, , \bar{\tau} \, , b \right\}$$

$$\mathbf{B}_{387} = \left\{ e \, , d \, , \nu_\mu \, , c \, , \bar{c} \, , \bar{s} \, , \nu_\tau \, , \bar{\nu}_\tau \right\}$$

$$\mathbf{B}_{388} = \left\{ e \, , d \, , \mu \, , \bar{\nu}_\mu \, , \bar{\mu} \, , s \, , \nu_\tau \, , \bar{\nu}_\tau \right\}$$

$$\mathbf{B}_{389} = \left\{ e \, , d \, , \mu \, , \bar{\nu}_\mu \, , c \, , \bar{s} \, , \bar{\tau} \, , b \right\}$$

$$\mathbf{B}_{390} = \left\{ e \, , d \, , \mu \, , \bar{\mu} \, , \bar{c} \, , \bar{s} \, , \bar{t} \, , \bar{b} \right\}$$

$$\mathbf{B}_{391} = \left\{ e \, , d \, , \mu \, , c \, , s \, , \bar{c} \, , \tau \, , t \right\}$$

$$\mathbf{B}_{392} = \left\{ e, d, \bar{\nu}_\mu, \bar{c}, \nu_\tau, \tau, \bar{\tau}, \bar{t} \right\}$$

$$\mathbf{B}_{393} = \left\{ e, d, \bar{\nu}_\mu, \bar{c}, \bar{\nu}_\tau, t, b, \bar{b} \right\}$$

$$\mathbf{B}_{394} = \left\{ e, d, \bar{\mu}, c, \nu_\tau, \tau, b, \bar{b} \right\}$$

$$\mathbf{B}_{395} = \left\{ e, d, \bar{\mu}, c, \bar{\nu}_\tau, \bar{\tau}, t, \bar{t} \right\}$$

$$\mathbf{B}_{396} = \left\{ e, d, s, \bar{s}, \nu_\tau, t, b, \bar{t} \right\}$$

$$\mathbf{B}_{397} = \left\{ e, d, s, \bar{s}, \tau, \bar{\nu}_\tau, \bar{\tau}, \bar{b} \right\}$$

$$\mathbf{B}_{398} = \left\{ e, \bar{u}, \nu_\mu, \mu, \bar{\nu}_\mu, c, \nu_\tau, \tau \right\}$$

$$\mathbf{B}_{399} = \left\{ e, \bar{u}, \nu_\mu, \mu, \bar{\mu}, \bar{c}, \bar{\nu}_\tau, t \right\}$$

$$\mathbf{B}_{400} = \left\{ e, \bar{u}, \nu_\mu, \bar{\nu}_\mu, \bar{c}, \bar{s}, b, \bar{t} \right\}$$

$$\mathbf{B}_{401} = \left\{ e, \bar{u}, \nu_\mu, \bar{\mu}, c, \bar{s}, \bar{\tau}, \bar{b} \right\}$$

$$\mathbf{B}_{402} = \left\{ e, \bar{u}, \nu_\mu, s, \nu_\tau, \bar{\nu}_\tau, b, \bar{b} \right\}$$

$$\mathbf{B}_{403} = \left\{ e, \bar{u}, \nu_\mu, s, \tau, \bar{\tau}, t, \bar{t} \right\}$$

$$\mathbf{B}_{404} = \left\{ e, \bar{u}, \mu, \bar{\nu}_\mu, s, \bar{c}, \bar{\tau}, \bar{b} \right\}$$

$$\mathbf{B}_{405} = \left\{ e, \bar{u}, \mu, \bar{\mu}, c, s, b, \bar{t} \right\}$$

$$\mathbf{B}_{406} = \left\{ e, \bar{u}, \mu, \bar{s}, \nu_\tau, \bar{\nu}_\tau, \bar{\tau}, \bar{t} \right\}$$

$$\mathbf{B}_{407} = \left\{ e, \bar{u}, \mu, \bar{s}, \tau, t, b, \bar{b} \right\}$$

$$\mathbf{B}_{408} = \left\{ e, \bar{u}, \bar{\nu}_\mu, \bar{\mu}, \nu_\tau, \bar{\tau}, t, b \right\}$$

$$\mathbf{B}_{409} = \left\{ e, \bar{u}, \bar{\nu}_\mu, \bar{\mu}, \tau, \bar{\nu}_\tau, \bar{t}, \bar{b} \right\}$$

$$\mathbf{B}_{410} = \left\{ e, \bar{u}, \bar{\nu}_\mu, c, s, \bar{s}, \bar{\nu}_\tau, t \right\}$$

$$\mathbf{B}_{411} = \left\{ e, \bar{u}, \bar{\mu}, s, \bar{c}, \bar{s}, \nu_\tau, \tau \right\}$$

$$\mathbf{B}_{412} = \left\{ e, \bar{u}, c, \bar{c}, \nu_\tau, t, \bar{t}, \bar{b} \right\}$$

$$\mathbf{B}_{413} = \left\{ e, \bar{u}, c, \bar{c}, \tau, \bar{\nu}_\tau, \bar{\tau}, b \right\}$$

$$\mathbf{B}_{414} = \left\{ e, \bar{d}, \nu_\mu, \mu, \bar{\mu}, s, \tau, \bar{b} \right\}$$

$$\mathbf{B}_{415} = \left\{ e, \bar{d}, \nu_\mu, \mu, c, \bar{s}, t, \bar{t} \right\}$$

$$\mathbf{B}_{416} = \left\{ e, \bar{d}, \nu_\mu, \bar{\nu}_\mu, \bar{\mu}, c, \bar{\nu}_\tau, b \right\}$$

$$\mathbf{B}_{417} = \left\{ e, \bar{d}, \nu_\mu, \bar{\nu}_\mu, s, \bar{s}, \nu_\tau, \bar{\tau} \right\}$$

$$\mathbf{B}_{418} = \left\{ e, \bar{d}, \nu_\mu, \bar{c}, \nu_\tau, \tau, t, b \right\}$$

$$\mathbf{B}_{419} = \left\{ e, \bar{d}, \nu_\mu, \bar{c}, \bar{\nu}_\tau, \bar{\tau}, \bar{t}, \bar{b} \right\}$$

$$\mathbf{B}_{420} = \left\{ e, \bar{d}, \mu, \bar{\nu}_\mu, \nu_\tau, b, \bar{t}, \bar{b} \right\}$$

$$\mathbf{B}_{421} = \left\{ e, \bar{d}, \mu, \bar{\nu}_\mu, \tau, \bar{\nu}_\tau, \bar{\tau}, t \right\}$$

$$\mathbf{B}_{422} = \left\{ e, \bar{d}, \mu, \bar{\mu}, c, \bar{c}, \nu_\tau, \bar{\tau} \right\}$$

$$\mathbf{B}_{423} = \left\{ e, \bar{d}, \mu, s, \bar{c}, \bar{s}, \bar{\nu}_\tau, b \right\}$$

$$\mathbf{B}_{424} = \left\{ e, \bar{d}, \bar{\nu}_\mu, \bar{\mu}, s, \bar{c}, t, \bar{t} \right\}$$

$$\mathbf{B}_{425} = \left\{ e, \bar{d}, \bar{\nu}_\mu, c, \bar{c}, \bar{s}, \tau, \bar{b} \right\}$$

$$\mathbf{B}_{426} = \left\{ e, \bar{d}, \bar{\mu}, \bar{s}, \nu_\tau, \bar{\nu}_\tau, t, \bar{b} \right\}$$

$$\mathbf{B}_{427} = \left\{ e, \bar{d}, \bar{\mu}, \bar{s}, \tau, \bar{\tau}, b, \bar{t} \right\}$$

$$\mathbf{B}_{428} = \left\{ e, \bar{d}, c, s, \nu_\tau, \tau, \bar{\nu}_\tau, \bar{t} \right\}$$

$$\mathbf{B}_{429} = \left\{ e, \bar{d}, c, s, \bar{\tau}, t, b, \bar{b} \right\}$$

$$\mathbf{B}_{430} = \left\{ \bar{\nu}_e, \bar{e}, u, d, \nu_\mu, c, s, \bar{c} \right\}$$

$$\mathbf{B}_{431} = \left\{ \bar{\nu}_e, \bar{e}, u, d, \mu, \bar{\nu}_\mu, \bar{\mu}, \bar{s} \right\}$$

$$\mathbf{B}_{432} = \left\{ \bar{\nu}_e, \bar{e}, u, d, \nu_\tau, \tau, \bar{\tau}, \bar{b} \right\}$$

$$\mathbf{B}_{433} = \left\{ \bar{\nu}_e, \bar{e}, u, d, \bar{\nu}_\tau, t, b, \bar{t} \right\}$$

$$\mathbf{B}_{434} = \left\{ \bar{\nu}_e, \bar{e}, u, \bar{u}, \nu_\mu, \bar{s}, b, \bar{b} \right\}$$

$$\mathbf{B}_{435} = \left\{ \bar{\nu}_e, \bar{e}, u, \bar{u}, \mu, s, \bar{\tau}, \bar{t} \right\}$$

$$\mathbf{B}_{436} = \left\{ \bar{\nu}_e, \bar{e}, u, \bar{u}, \bar{\nu}_\mu, c, \nu_\tau, t \right\}$$

$$\mathbf{B}_{437} = \left\{ \bar{\nu}_e, \bar{e}, u, \bar{u}, \bar{\mu}, \bar{c}, \tau, \bar{\nu}_\tau \right\}$$

$$\mathbf{B}_{438} = \left\{ \bar{\nu}_e, \bar{e}, u, \bar{d}, \nu_\mu, \bar{\nu}_\mu, \bar{\nu}_\tau, \bar{\tau} \right\}$$

$$\mathbf{B}_{439} = \left\{ \bar{\nu}_e, \bar{e}, u, \bar{d}, \mu, \bar{c}, \nu_\tau, b \right\}$$

$$\mathbf{B}_{440} = \left\{ \bar{\nu}_e, \bar{e}, u, \bar{d}, \bar{\mu}, s, t, \bar{b} \right\}$$

$$\mathbf{B}_{441} = \left\{ \bar{\nu}_e, \bar{e}, u, \bar{d}, c, \bar{s}, \tau, \bar{t} \right\}$$

$$\mathbf{B}_{442} = \left\{ \bar{\nu}_e, \bar{e}, d, \bar{u}, \nu_\mu, \bar{\nu}_\mu, \tau, \bar{t} \right\}$$

$$\mathbf{B}_{443} = \left\{ \bar{\nu}_e, \bar{e}, d, \bar{u}, \mu, \bar{c}, t, \bar{b} \right\}$$

$$\mathbf{B}_{444} = \left\{ \bar{\nu}_e, \bar{e}, d, \bar{u}, \bar{\mu}, s, \nu_\tau, b \right\}$$

$$\mathbf{B}_{445} = \left\{ \bar{\nu}_e, \bar{e}, d, \bar{u}, c, \bar{s}, \bar{\nu}_\tau, \bar{\tau} \right\}$$

$$\mathbf{B}_{446} = \left\{ \bar{\nu}_e, \bar{e}, d, \bar{d}, \nu_\mu, \bar{s}, \nu_\tau, t \right\}$$

$$\mathbf{B}_{447} = \left\{ \bar{\nu}_e, \bar{e}, d, \bar{d}, \mu, s, \tau, \bar{\nu}_\tau \right\}$$

$$\mathbf{B_{448}} = \left\{ \bar{\nu}_e, \bar{e}, d, \bar{d}, \bar{\nu}_\mu, c, b, \bar{b} \right\}$$

$$\mathbf{B_{449}} = \left\{ \bar{\nu}_e, \bar{e}, d, \bar{d}, \bar{\mu}, \bar{c}, \bar{\tau}, \bar{t} \right\}$$

$$\mathbf{B_{450}} = \left\{ \bar{\nu}_e, \bar{e}, \bar{u}, \bar{d}, \nu_\mu, \mu, \bar{\mu}, c \right\}$$

$$\mathbf{B_{451}} = \left\{ \bar{\nu}_e, \bar{e}, \bar{u}, \bar{d}, \bar{\nu}_\mu, s, \bar{c}, \bar{s} \right\}$$

$$\mathbf{B_{452}} = \left\{ \bar{\nu}_e, \bar{e}, \bar{u}, \bar{d}, \nu_\tau, \bar{\nu}_\tau, \bar{t}, \bar{b} \right\}$$

$$\mathbf{B_{453}} = \left\{ \bar{\nu}_e, \bar{e}, \bar{u}, \bar{d}, \tau, \bar{\tau}, t, b \right\}$$

$$\mathbf{B_{454}} = \left\{ \bar{\nu}_e, \bar{e}, \nu_\mu, \mu, \bar{\nu}_\mu, s, \nu_\tau, \bar{b} \right\}$$

$$\mathbf{B_{455}} = \left\{ \bar{\nu}_e, \bar{e}, \nu_\mu, \mu, \bar{c}, \bar{s}, \bar{\nu}_\tau, \bar{t} \right\}$$

$$\mathbf{B_{456}} = \left\{ \bar{\nu}_e, \bar{e}, \nu_\mu, \bar{\nu}_\mu, \bar{\mu}, \bar{c}, t, b \right\}$$

$$\mathbf{B_{457}} = \left\{ \bar{\nu}_e, \bar{e}, \nu_\mu, \bar{\mu}, s, \bar{s}, \tau, \bar{\tau} \right\}$$

$$\mathbf{B_{458}} = \left\{ \bar{\nu}_e, \bar{e}, \nu_\mu, c, \nu_\tau, \tau, \bar{\nu}_\tau, b \right\}$$

$$\mathbf{B_{459}} = \left\{ \bar{\nu}_e, \bar{e}, \nu_\mu, c, \bar{\tau}, t, \bar{t}, \bar{b} \right\}$$

$$\mathbf{B_{460}} = \left\{ \bar{\nu}_e, \bar{e}, \mu, \bar{\nu}_\mu, c, \bar{c}, \tau, \bar{\tau} \right\}$$

$$\mathbf{B_{461}} = \left\{ \bar{\nu}_e, \bar{e}, \mu, \bar{\mu}, \nu_\tau, \bar{\nu}_\tau, \bar{\tau}, t \right\}$$

$$\mathbf{B}_{462} = \left\{ \bar{\nu}_e, \bar{e}, \mu, \bar{\mu}, \tau, b, \bar{t}, \bar{b} \right\}$$

$$\mathbf{B}_{463} = \left\{ \bar{\nu}_e, \bar{e}, \mu, c, s, \bar{s}, t, b \right\}$$

$$\mathbf{B}_{464} = \left\{ \bar{\nu}_e, \bar{e}, \bar{\nu}_\mu, \bar{\mu}, c, s, \bar{\nu}_\tau, \bar{t} \right\}$$

$$\mathbf{B}_{465} = \left\{ \bar{\nu}_e, \bar{e}, \bar{\nu}_\mu, \bar{s}, \nu_\tau, \bar{\tau}, b, \bar{t} \right\}$$

$$\mathbf{B}_{466} = \left\{ \bar{\nu}_e, \bar{e}, \bar{\nu}_\mu, \bar{s}, \tau, \bar{\nu}_\tau, t, \bar{b} \right\}$$

$$\mathbf{B}_{467} = \left\{ \bar{\nu}_e, \bar{e}, \bar{\mu}, c, \bar{c}, \bar{s}, \nu_\tau, \bar{b} \right\}$$

$$\mathbf{B}_{468} = \left\{ \bar{\nu}_e, \bar{e}, s, \bar{c}, \nu_\tau, \tau, t, \bar{t} \right\}$$

$$\mathbf{B}_{469} = \left\{ \bar{\nu}_e, \bar{e}, s, \bar{c}, \bar{\nu}_\tau, \bar{\tau}, b, \bar{b} \right\}$$

$$\mathbf{B}_{470} = \left\{ \bar{\nu}_e, u, d, \bar{u}, \nu_\mu, \mu, \nu_\tau, \bar{\nu}_\tau \right\}$$

$$\mathbf{B}_{471} = \left\{ \bar{\nu}_e, u, d, \bar{u}, \bar{\nu}_\mu, \bar{c}, \bar{\tau}, b \right\}$$

$$\mathbf{B}_{472} = \left\{ \bar{\nu}_e, u, d, \bar{u}, \bar{\mu}, c, \bar{t}, \bar{b} \right\}$$

$$\mathbf{B}_{473} = \left\{ \bar{\nu}_e, u, d, \bar{u}, s, \bar{s}, \tau, t \right\}$$

$$\mathbf{B}_{474} = \left\{ \bar{\nu}_e, u, d, \bar{d}, \nu_\mu, \bar{\mu}, \tau, b \right\}$$

$$\mathbf{B}_{475} = \left\{ \bar{\nu}_e, u, d, \bar{d}, \mu, c, \bar{\tau}, t \right\}$$

$$\mathbf{B}_{476} = \left\{ \bar{\nu}_e, u, d, \bar{d}, \bar{\nu}_\mu, s, \nu_\tau, \bar{t} \right\}$$

$$\mathbf{B}_{477} = \left\{ \bar{\nu}_e, u, d, \bar{d}, \bar{c}, \bar{s}, \bar{\nu}_\tau, \bar{b} \right\}$$

$$\mathbf{B}_{478} = \left\{ \bar{\nu}_e, u, \bar{u}, \bar{d}, \nu_\mu, \bar{c}, t, \bar{t} \right\}$$

$$\mathbf{B}_{479} = \left\{ \bar{\nu}_e, u, \bar{u}, \bar{d}, \mu, \bar{\nu}_\mu, \tau, \bar{b} \right\}$$

$$\mathbf{B}_{480} = \left\{ \bar{\nu}_e, u, \bar{u}, \bar{d}, \bar{\mu}, \bar{s}, \nu_\tau, \bar{\tau} \right\}$$

$$\mathbf{B}_{481} = \left\{ \bar{\nu}_e, u, \bar{u}, \bar{d}, c, s, \bar{\nu}_\tau, b \right\}$$

$$\mathbf{B}_{482} = \left\{ \bar{\nu}_e, u, \nu_\mu, \mu, \bar{\nu}_\mu, c, b, \bar{t} \right\}$$

$$\mathbf{B}_{483} = \left\{ \bar{\nu}_e, u, \nu_\mu, \mu, \bar{\mu}, \bar{c}, \bar{\tau}, \bar{b} \right\}$$

$$\mathbf{B}_{484} = \left\{ \bar{\nu}_e, u, \nu_\mu, \bar{\nu}_\mu, \bar{c}, \bar{s}, \nu_\tau, \tau \right\}$$

$$\mathbf{B}_{485} = \left\{ \bar{\nu}_e, u, \nu_\mu, \bar{\mu}, c, \bar{s}, \bar{\nu}_\tau, t \right\}$$

$$\mathbf{B}_{486} = \left\{ \bar{\nu}_e, u, \nu_\mu, s, \nu_\tau, \bar{\tau}, t, b \right\}$$

$$\mathbf{B}_{487} = \left\{ \bar{\nu}_e, u, \nu_\mu, s, \tau, \bar{\nu}_\tau, \bar{t}, \bar{b} \right\}$$

$$\mathbf{B}_{488} = \left\{ \bar{\nu}_e, u, \mu, \bar{\nu}_\mu, s, \bar{c}, \bar{\nu}_\tau, t \right\}$$

$$\mathbf{B}_{489} = \left\{ \bar{\nu}_e, u, \mu, \bar{\mu}, c, s, \nu_\tau, \tau \right\}$$

$$\mathbf{B}_{490} = \left\{ \bar{\nu}_e, u, \mu, \bar{s}, \nu_\tau, t, \bar{t}, \bar{b} \right\}$$

$$\mathbf{B}_{491} = \left\{ \bar{\nu}_e, u, \mu, \bar{s}, \tau, \bar{\nu}_\tau, \bar{\tau}, b \right\}$$

$$\mathbf{B}_{492} = \left\{ \bar{\nu}_e, u, \bar{\nu}_\mu, \bar{\mu}, \nu_\tau, \bar{\nu}_\tau, b, \bar{b} \right\}$$

$$\mathbf{B}_{493} = \left\{ \bar{\nu}_e, u, \bar{\nu}_\mu, \bar{\mu}, \tau, \bar{\tau}, t, \bar{t} \right\}$$

$$\mathbf{B}_{494} = \left\{ \bar{\nu}_e, u, \bar{\nu}_\mu, c, s, \bar{s}, \bar{\tau}, \bar{b} \right\}$$

$$\mathbf{B}_{495} = \left\{ \bar{\nu}_e, u, \bar{\mu}, s, \bar{c}, \bar{s}, b, \bar{t} \right\}$$

$$\mathbf{B}_{496} = \left\{ \bar{\nu}_e, u, c, \bar{c}, \nu_\tau, \bar{\nu}_\tau, \bar{\tau}, \bar{t} \right\}$$

$$\mathbf{B}_{497} = \left\{ \bar{\nu}_e, u, c, \bar{c}, \tau, t, b, \bar{b} \right\}$$

$$\mathbf{B}_{498} = \left\{ \bar{\nu}_e, d, \bar{u}, \bar{d}, \nu_\mu, s, \bar{\tau}, \bar{b} \right\}$$

$$\mathbf{B}_{499} = \left\{ \bar{\nu}_e, d, \bar{u}, \bar{d}, \mu, \bar{s}, b, \bar{t} \right\}$$

$$\mathbf{B}_{500} = \left\{ \bar{\nu}_e, d, \bar{u}, \bar{d}, \bar{\nu}_\mu, \bar{\mu}, \bar{\nu}_\tau, t \right\}$$

$$\mathbf{B}_{501} = \left\{ \bar{\nu}_e, d, \bar{u}, \bar{d}, c, \bar{c}, \nu_\tau, \tau \right\}$$

$$\mathbf{B}_{502} = \left\{ \bar{\nu}_e, d, \nu_\mu, \mu, \bar{\mu}, s, t, \bar{t} \right\}$$

$$\mathbf{B}_{503} = \left\{ \bar{\nu}_e, d, \nu_\mu, \mu, c, \bar{s}, \tau, \bar{b} \right\}$$

$$\mathbf{B}_{504} = \left\{ \bar{\nu}_e, d, \nu_\mu, \bar{\nu}_\mu, \bar{\mu}, c, \nu_\tau, \bar{\tau} \right\}$$

$$\mathbf{B}_{505} = \left\{ \bar{\nu}_e, d, \nu_\mu, \bar{\nu}_\mu, s, \bar{s}, \bar{\nu}_\tau, b \right\}$$

$$\mathbf{B}_{506} = \left\{ \bar{\nu}_e, d, \nu_\mu, \bar{c}, \nu_\tau, b, \bar{t}, \bar{b} \right\}$$

$$\mathbf{B}_{507} = \left\{ \bar{\nu}_e, d, \nu_\mu, \bar{c}, \tau, \bar{\nu}_\tau, \bar{\tau}, t \right\}$$

$$\mathbf{B}_{508} = \left\{ \bar{\nu}_e, d, \mu, \bar{\nu}_\mu, \nu_\tau, \tau, t, b \right\}$$

$$\mathbf{B}_{509} = \left\{ \bar{\nu}_e, d, \mu, \bar{\nu}_\mu, \bar{\nu}_\tau, \bar{\tau}, \bar{t}, \bar{b} \right\}$$

$$\mathbf{B}_{510} = \left\{ \bar{\nu}_e, d, \mu, \bar{\mu}, c, \bar{c}, \bar{\nu}_\tau, b \right\}$$

$$\mathbf{B}_{511} = \left\{ \bar{\nu}_e, d, \mu, s, \bar{c}, \bar{s}, \nu_\tau, \bar{\tau} \right\}$$

$$\mathbf{B}_{512} = \left\{ \bar{\nu}_e, d, \bar{\nu}_\mu, \bar{\mu}, s, \bar{c}, \tau, \bar{b} \right\}$$

$$\mathbf{B}_{513} = \left\{ \bar{\nu}_e, d, \bar{\nu}_\mu, c, \bar{c}, \bar{s}, t, \bar{t} \right\}$$

$$\mathbf{B}_{514} = \left\{ \bar{\nu}_e, d, \bar{\mu}, \bar{s}, \nu_\tau, \tau, \bar{\nu}_\tau, \bar{t} \right\}$$

$$\mathbf{B}_{515} = \left\{ \bar{\nu}_e, d, \bar{\mu}, \bar{s}, \bar{\tau}, t, b, \bar{b} \right\}$$

$$\mathbf{B}_{516} = \left\{ \bar{\nu}_e, d, c, s, \nu_\tau, \bar{\nu}_\tau, t, \bar{b} \right\}$$

$$\mathbf{B}_{517} = \left\{ \bar{\nu}_e, d, c, s, \tau, \bar{\tau}, b, \bar{t} \right\}$$

$$\mathbf{B_{518}} = \left\{ \bar{\nu}_e, \bar{u}, \nu_\mu, \mu, \bar{\nu}_\mu, \bar{s}, \bar{\tau}, t \right\}$$

$$\mathbf{B_{519}} = \left\{ \bar{\nu}_e, \bar{u}, \nu_\mu, \mu, s, \bar{c}, \tau, b \right\}$$

$$\mathbf{B_{520}} = \left\{ \bar{\nu}_e, \bar{u}, \nu_\mu, \bar{\nu}_\mu, c, \bar{c}, \bar{\nu}_\tau, \bar{b} \right\}$$

$$\mathbf{B_{521}} = \left\{ \bar{\nu}_e, \bar{u}, \nu_\mu, \bar{\mu}, \nu_\tau, \tau, t, \bar{b} \right\}$$

$$\mathbf{B_{522}} = \left\{ \bar{\nu}_e, \bar{u}, \nu_\mu, \bar{\mu}, \bar{\nu}_\tau, \bar{\tau}, b, \bar{t} \right\}$$

$$\mathbf{B_{523}} = \left\{ \bar{\nu}_e, \bar{u}, \nu_\mu, c, s, \bar{s}, \nu_\tau, \bar{t} \right\}$$

$$\mathbf{B_{524}} = \left\{ \bar{\nu}_e, \bar{u}, \mu, \bar{\nu}_\mu, \bar{\mu}, \bar{c}, \nu_\tau, \bar{t} \right\}$$

$$\mathbf{B_{525}} = \left\{ \bar{\nu}_e, \bar{u}, \mu, \bar{\mu}, s, \bar{s}, \bar{\nu}_\tau, \bar{b} \right\}$$

$$\mathbf{B_{526}} = \left\{ \bar{\nu}_e, \bar{u}, \mu, c, \nu_\tau, \bar{\tau}, b, \bar{b} \right\}$$

$$\mathbf{B_{527}} = \left\{ \bar{\nu}_e, \bar{u}, \mu, c, \tau, \bar{\nu}_\tau, t, \bar{t} \right\}$$

$$\mathbf{B_{528}} = \left\{ \bar{\nu}_e, \bar{u}, \bar{\nu}_\mu, \bar{\mu}, c, \bar{s}, \tau, b \right\}$$

$$\mathbf{B_{529}} = \left\{ \bar{\nu}_e, \bar{u}, \bar{\nu}_\mu, s, \nu_\tau, \tau, \bar{\nu}_\tau, \bar{\tau} \right\}$$

$$\mathbf{B_{530}} = \left\{ \bar{\nu}_e, \bar{u}, \bar{\nu}_\mu, s, t, b, \bar{t}, \bar{b} \right\}$$

$$\mathbf{B_{531}} = \left\{ \bar{\nu}_e, \bar{u}, \bar{\mu}, c, s, \bar{c}, \bar{\tau}, t \right\}$$

$$\mathbf{B}_{532} = \left\{ \bar{\nu}_e, \bar{u}, \bar{c}, \bar{s}, \nu_\tau, \bar{\nu}_\tau, t, b \right\}$$

$$\mathbf{B}_{533} = \left\{ \bar{\nu}_e, \bar{u}, \bar{c}, \bar{s}, \tau, \bar{\tau}, \bar{t}, \bar{b} \right\}$$

$$\mathbf{B}_{534} = \left\{ \bar{\nu}_e, \bar{d}, \nu_\mu, \mu, \nu_\tau, \tau, \bar{\tau}, \bar{t} \right\}$$

$$\mathbf{B}_{535} = \left\{ \bar{\nu}_e, \bar{d}, \nu_\mu, \mu, \bar{\nu}_\tau, t, b, \bar{b} \right\}$$

$$\mathbf{B}_{536} = \left\{ \bar{\nu}_e, \bar{d}, \nu_\mu, \bar{\nu}_\mu, \bar{\mu}, \bar{s}, \bar{t}, \bar{b} \right\}$$

$$\mathbf{B}_{537} = \left\{ \bar{\nu}_e, \bar{d}, \nu_\mu, \bar{\nu}_\mu, c, s, \tau, t \right\}$$

$$\mathbf{B}_{538} = \left\{ \bar{\nu}_e, \bar{d}, \nu_\mu, \bar{\mu}, s, \bar{c}, \nu_\tau, \bar{\nu}_\tau \right\}$$

$$\mathbf{B}_{539} = \left\{ \bar{\nu}_e, \bar{d}, \nu_\mu, c, \bar{c}, \bar{s}, \bar{\tau}, b \right\}$$

$$\mathbf{B}_{540} = \left\{ \bar{\nu}_e, \bar{d}, \mu, \bar{\nu}_\mu, \bar{\mu}, s, \bar{\tau}, b \right\}$$

$$\mathbf{B}_{541} = \left\{ \bar{\nu}_e, \bar{d}, \mu, \bar{\nu}_\mu, c, \bar{s}, \nu_\tau, \bar{\nu}_\tau \right\}$$

$$\mathbf{B}_{542} = \left\{ \bar{\nu}_e, \bar{d}, \mu, \bar{\mu}, \bar{c}, \bar{s}, \tau, t \right\}$$

$$\mathbf{B}_{543} = \left\{ \bar{\nu}_e, \bar{d}, \mu, c, s, \bar{c}, \bar{t}, \bar{b} \right\}$$

$$\mathbf{B}_{544} = \left\{ \bar{\nu}_e, \bar{d}, \bar{\nu}_\mu, \bar{c}, \nu_\tau, \bar{\tau}, t, \bar{b} \right\}$$

$$\mathbf{B}_{545} = \left\{ \bar{\nu}_e, \bar{d}, \bar{\nu}_\mu, \bar{c}, \tau, \bar{\nu}_\tau, b, \bar{t} \right\}$$

$$\mathbf{B_{546}} = \left\{ \bar{\nu}_e, \bar{d}, \bar{\mu}, c, \nu_\tau, t, b, \bar{t} \right\}$$

$$\mathbf{B_{547}} = \left\{ \bar{\nu}_e, \bar{d}, \bar{\mu}, c, \tau, \bar{\nu}_\tau, \bar{\tau}, \bar{b} \right\}$$

$$\mathbf{B_{548}} = \left\{ \bar{\nu}_e, \bar{d}, s, \bar{s}, \nu_\tau, \tau, b, \bar{b} \right\}$$

$$\mathbf{B_{549}} = \left\{ \bar{\nu}_e, \bar{d}, s, \bar{s}, \bar{\nu}_\tau, \bar{\tau}, t, \bar{t} \right\}$$

$$\mathbf{B_{550}} = \left\{ \bar{e}, u, d, \bar{u}, \nu_\mu, \bar{\mu}, \bar{\tau}, t \right\}$$

$$\mathbf{B_{551}} = \left\{ \bar{e}, u, d, \bar{u}, \mu, c, \tau, b \right\}$$

$$\mathbf{B_{552}} = \left\{ \bar{e}, u, d, \bar{u}, \bar{\nu}_\mu, s, \bar{\nu}_\tau, \bar{b} \right\}$$

$$\mathbf{B_{553}} = \left\{ \bar{e}, u, d, \bar{u}, \bar{c}, \bar{s}, \nu_\tau, \bar{t} \right\}$$

$$\mathbf{B_{554}} = \left\{ \bar{e}, u, d, \bar{d}, \nu_\mu, \mu, \bar{t}, \bar{b} \right\}$$

$$\mathbf{B_{555}} = \left\{ \bar{e}, u, d, \bar{d}, \bar{\nu}_\mu, \bar{c}, \tau, t \right\}$$

$$\mathbf{B_{556}} = \left\{ \bar{e}, u, d, \bar{d}, \bar{\mu}, c, \nu_\tau, \bar{\nu}_\tau \right\}$$

$$\mathbf{B_{557}} = \left\{ \bar{e}, u, d, \bar{d}, s, \bar{s}, \bar{\tau}, b \right\}$$

$$\mathbf{B_{558}} = \left\{ \bar{e}, u, \bar{u}, \bar{d}, \nu_\mu, s, \nu_\tau, \tau \right\}$$

$$\mathbf{B_{559}} = \left\{ \bar{e}, u, \bar{u}, \bar{d}, \mu, \bar{s}, \bar{\nu}_\tau, t \right\}$$

$$\mathbf{B}_{560} = \left\{ \bar{e}, u, \bar{u}, \bar{d}, \bar{\nu}_\mu, \bar{\mu}, b, \bar{t} \right\}$$

$$\mathbf{B}_{561} = \left\{ \bar{e}, u, \bar{u}, \bar{d}, c, \bar{c}, \bar{\tau}, \bar{b} \right\}$$

$$\mathbf{B}_{562} = \left\{ \bar{e}, u, \nu_\mu, \mu, \bar{\mu}, s, \bar{\nu}_\tau, b \right\}$$

$$\mathbf{B}_{563} = \left\{ \bar{e}, u, \nu_\mu, \mu, c, \bar{s}, \nu_\tau, \bar{\tau} \right\}$$

$$\mathbf{B}_{564} = \left\{ \bar{e}, u, \nu_\mu, \bar{\nu}_\mu, \bar{\mu}, c, \tau, \bar{b} \right\}$$

$$\mathbf{B}_{565} = \left\{ \bar{e}, u, \nu_\mu, \bar{\nu}_\mu, s, \bar{s}, t, \bar{t} \right\}$$

$$\mathbf{B}_{566} = \left\{ \bar{e}, u, \nu_\mu, \bar{c}, \nu_\tau, \bar{\nu}_\tau, t, \bar{b} \right\}$$

$$\mathbf{B}_{567} = \left\{ \bar{e}, u, \nu_\mu, \bar{c}, \tau, \bar{\tau}, b, \bar{t} \right\}$$

$$\mathbf{B}_{568} = \left\{ \bar{e}, u, \mu, \bar{\nu}_\mu, \nu_\tau, \tau, \bar{\nu}_\tau, \bar{t} \right\}$$

$$\mathbf{B}_{569} = \left\{ \bar{e}, u, \mu, \bar{\nu}_\mu, \bar{\tau}, t, b, \bar{b} \right\}$$

$$\mathbf{B}_{570} = \left\{ \bar{e}, u, \mu, \bar{\mu}, c, \bar{c}, t, \bar{t} \right\}$$

$$\mathbf{B}_{571} = \left\{ \bar{e}, u, \mu, s, \bar{c}, \bar{s}, \tau, \bar{b} \right\}$$

$$\mathbf{B}_{572} = \left\{ \bar{e}, u, \bar{\nu}_\mu, \bar{\mu}, s, \bar{c}, \nu_\tau, \bar{\tau} \right\}$$

$$\mathbf{B}_{573} = \left\{ \bar{e}, u, \bar{\nu}_\mu, c, \bar{c}, \bar{s}, \bar{\nu}_\tau, b \right\}$$

$$\mathbf{B}_{574} = \left\{ \bar{e}, u, \bar{\mu}, \bar{s}, \nu_\tau, \tau, t, b \right\}$$

$$\mathbf{B}_{575} = \left\{ \bar{e}, u, \bar{\mu}, \bar{s}, \bar{\nu}_\tau, \bar{\tau}, \bar{t}, \bar{b} \right\}$$

$$\mathbf{B}_{576} = \left\{ \bar{e}, u, c, s, \nu_\tau, b, \bar{t}, \bar{b} \right\}$$

$$\mathbf{B}_{577} = \left\{ \bar{e}, u, c, s, \tau, \bar{\nu}_\tau, \bar{\tau}, t \right\}$$

$$\mathbf{B}_{578} = \left\{ \bar{e}, d, \bar{u}, \bar{d}, \nu_\mu, \bar{c}, \bar{\nu}_\tau, b \right\}$$

$$\mathbf{B}_{579} = \left\{ \bar{e}, d, \bar{u}, \bar{d}, \mu, \bar{\nu}_\mu, \nu_\tau, \bar{\tau} \right\}$$

$$\mathbf{B}_{580} = \left\{ \bar{e}, d, \bar{u}, \bar{d}, \bar{\mu}, \bar{s}, \tau, \bar{b} \right\}$$

$$\mathbf{B}_{581} = \left\{ \bar{e}, d, \bar{u}, \bar{d}, c, s, t, \bar{t} \right\}$$

$$\mathbf{B}_{582} = \left\{ \bar{e}, d, \nu_\mu, \mu, \bar{\nu}_\mu, c, \bar{\nu}_\tau, t \right\}$$

$$\mathbf{B}_{583} = \left\{ \bar{e}, d, \nu_\mu, \mu, \bar{\mu}, \bar{c}, \nu_\tau, \tau \right\}$$

$$\mathbf{B}_{584} = \left\{ \bar{e}, d, \nu_\mu, \bar{\nu}_\mu, \bar{c}, \bar{s}, \bar{\tau}, \bar{b} \right\}$$

$$\mathbf{B}_{585} = \left\{ \bar{e}, d, \nu_\mu, \bar{\mu}, c, \bar{s}, b, \bar{t} \right\}$$

$$\mathbf{B}_{586} = \left\{ \bar{e}, d, \nu_\mu, s, \nu_\tau, \bar{\nu}_\tau, \bar{\tau}, \bar{t} \right\}$$

$$\mathbf{B}_{587} = \left\{ \bar{e}, d, \nu_\mu, s, \tau, t, b, \bar{b} \right\}$$

$$\mathbf{B}_{588} = \left\{ \bar{e}, d, \mu, \bar{\nu}_\mu, s, \bar{c}, b, \bar{t} \right\}$$

$$\mathbf{B}_{589} = \left\{ \bar{e}, d, \mu, \bar{\mu}, c, s, \bar{\tau}, \bar{b} \right\}$$

$$\mathbf{B}_{590} = \left\{ \bar{e}, d, \mu, \bar{s}, \nu_\tau, \bar{\nu}_\tau, b, \bar{b} \right\}$$

$$\mathbf{B}_{591} = \left\{ \bar{e}, d, \mu, \bar{s}, \tau, \bar{\tau}, t, \bar{t} \right\}$$

$$\mathbf{B}_{592} = \left\{ \bar{e}, d, \bar{\nu}_\mu, \bar{\mu}, \nu_\tau, t, \bar{t}, \bar{b} \right\}$$

$$\mathbf{B}_{593} = \left\{ \bar{e}, d, \bar{\nu}_\mu, \bar{\mu}, \tau, \bar{\nu}_\tau, \bar{\tau}, b \right\}$$

$$\mathbf{B}_{594} = \left\{ \bar{e}, d, \bar{\nu}_\mu, c, s, \bar{s}, \nu_\tau, \tau \right\}$$

$$\mathbf{B}_{595} = \left\{ \bar{e}, d, \bar{\mu}, s, \bar{c}, \bar{s}, \bar{\nu}_\tau, t \right\}$$

$$\mathbf{B}_{596} = \left\{ \bar{e}, d, c, \bar{c}, \nu_\tau, \bar{\tau}, t, b \right\}$$

$$\mathbf{B}_{597} = \left\{ \bar{e}, d, c, \bar{c}, \tau, \bar{\nu}_\tau, \bar{t}, \bar{b} \right\}$$

$$\mathbf{B}_{598} = \left\{ \bar{e}, \bar{u}, \nu_\mu, \mu, \nu_\tau, t, b, \bar{t} \right\}$$

$$\mathbf{B}_{599} = \left\{ \bar{e}, \bar{u}, \nu_\mu, \mu, \tau, \bar{\nu}_\tau, \bar{\tau}, \bar{b} \right\}$$

$$\mathbf{B}_{600} = \left\{ \bar{e}, \bar{u}, \nu_\mu, \bar{\nu}_\mu, \bar{\mu}, \bar{s}, \nu_\tau, \bar{\nu}_\tau \right\}$$

$$\mathbf{B}_{601} = \left\{ \bar{e}, \bar{u}, \nu_\mu, \bar{\nu}_\mu, c, s, \bar{\tau}, b \right\}$$

$$\mathbf{B}_{602} = \left\{ \bar{e}, \bar{u}, \nu_{\mu}, \bar{\mu}, s, \bar{c}, \bar{t}, \bar{b} \right\}$$

$$\mathbf{B}_{603} = \left\{ \bar{e}, \bar{u}, \nu_{\mu}, c, \bar{c}, \bar{s}, \tau, t \right\}$$

$$\mathbf{B}_{604} = \left\{ \bar{e}, \bar{u}, \mu, \bar{\nu}_{\mu}, \bar{\mu}, s, \tau, t \right\}$$

$$\mathbf{B}_{605} = \left\{ \bar{e}, \bar{u}, \mu, \bar{\nu}_{\mu}, c, \bar{s}, \bar{t}, \bar{b} \right\}$$

$$\mathbf{B}_{606} = \left\{ \bar{e}, \bar{u}, \mu, \bar{\mu}, \bar{c}, \bar{s}, \bar{\tau}, b \right\}$$

$$\mathbf{B}_{607} = \left\{ \bar{e}, \bar{u}, \mu, c, s, \bar{c}, \nu_{\tau}, \bar{\nu}_{\tau} \right\}$$

$$\mathbf{B}_{608} = \left\{ \bar{e}, \bar{u}, \bar{\nu}_{\mu}, \bar{c}, \nu_{\tau}, \tau, b, \bar{b} \right\}$$

$$\mathbf{B}_{609} = \left\{ \bar{e}, \bar{u}, \bar{\nu}_{\mu}, \bar{c}, \bar{\nu}_{\tau}, \bar{\tau}, t, \bar{t} \right\}$$

$$\mathbf{B}_{610} = \left\{ \bar{e}, \bar{u}, \bar{\mu}, c, \nu_{\tau}, \tau, \bar{\tau}, \bar{t} \right\}$$

$$\mathbf{B}_{611} = \left\{ \bar{e}, \bar{u}, \bar{\mu}, c, \bar{\nu}_{\tau}, t, b, \bar{b} \right\}$$

$$\mathbf{B}_{612} = \left\{ \bar{e}, \bar{u}, s, \bar{s}, \nu_{\tau}, \bar{\tau}, t, \bar{b} \right\}$$

$$\mathbf{B}_{613} = \left\{ \bar{e}, \bar{u}, s, \bar{s}, \tau, \bar{\nu}_{\tau}, b, \bar{t} \right\}$$

$$\mathbf{B}_{614} = \left\{ \bar{e}, \bar{d}, \nu_{\mu}, \mu, \bar{\nu}_{\mu}, \bar{s}, \tau, b \right\}$$

$$\mathbf{B}_{615} = \left\{ \bar{e}, \bar{d}, \nu_{\mu}, \mu, s, \bar{c}, \bar{\tau}, t \right\}$$

$$\mathbf{B}_{616} = \left\{ \bar{e}, \bar{d}, \nu_\mu, \bar{\nu}_\mu, c, \bar{c}, \nu_\tau, \bar{t} \right\}$$

$$\mathbf{B}_{617} = \left\{ \bar{e}, \bar{d}, \nu_\mu, \bar{\mu}, \nu_\tau, \bar{\tau}, b, \bar{b} \right\}$$

$$\mathbf{B}_{618} = \left\{ \bar{e}, \bar{d}, \nu_\mu, \bar{\mu}, \tau, \bar{\nu}_\tau, t, \bar{t} \right\}$$

$$\mathbf{B}_{619} = \left\{ \bar{e}, \bar{d}, \nu_\mu, c, s, \bar{s}, \bar{\nu}_\tau, \bar{b} \right\}$$

$$\mathbf{B}_{620} = \left\{ \bar{e}, \bar{d}, \mu, \bar{\nu}_\mu, \bar{\mu}, \bar{c}, \bar{\nu}_\tau, \bar{b} \right\}$$

$$\mathbf{B}_{621} = \left\{ \bar{e}, \bar{d}, \mu, \bar{\mu}, s, \bar{s}, \nu_\tau, \bar{t} \right\}$$

$$\mathbf{B}_{622} = \left\{ \bar{e}, \bar{d}, \mu, c, \nu_\tau, \tau, t, \bar{b} \right\}$$

$$\mathbf{B}_{623} = \left\{ \bar{e}, \bar{d}, \mu, c, \bar{\nu}_\tau, \bar{\tau}, b, \bar{t} \right\}$$

$$\mathbf{B}_{624} = \left\{ \bar{e}, \bar{d}, \bar{\nu}_\mu, \bar{\mu}, c, \bar{s}, \bar{\tau}, t \right\}$$

$$\mathbf{B}_{625} = \left\{ \bar{e}, \bar{d}, \bar{\nu}_\mu, s, \nu_\tau, \bar{\nu}_\tau, t, b \right\}$$

$$\mathbf{B}_{626} = \left\{ \bar{e}, \bar{d}, \bar{\nu}_\mu, s, \tau, \bar{\tau}, \bar{t}, \bar{b} \right\}$$

$$\mathbf{B}_{627} = \left\{ \bar{e}, \bar{d}, \bar{\mu}, c, s, \bar{c}, \tau, b \right\}$$

$$\mathbf{B}_{628} = \left\{ \bar{e}, \bar{d}, \bar{c}, \bar{s}, \nu_\tau, \tau, \bar{\nu}_\tau, \bar{\tau} \right\}$$

$$\mathbf{B}_{629} = \left\{ \bar{e}, \bar{d}, \bar{c}, \bar{s}, t, b, \bar{t}, \bar{b} \right\}$$

$$\mathbf{B}_{630} = \left\{ u, d, \bar{u}, \bar{d}, \nu_\mu, \bar{\nu}_\mu, c, \bar{s} \right\}$$

$$\mathbf{B}_{631} = \left\{ u, d, \bar{u}, \bar{d}, \mu, \bar{\mu}, s, \bar{c} \right\}$$

$$\mathbf{B}_{632} = \left\{ u, d, \bar{u}, \bar{d}, \nu_\tau, t, b, \bar{b} \right\}$$

$$\mathbf{B}_{633} = \left\{ u, d, \bar{u}, \bar{d}, \tau, \bar{\nu}_\tau, \bar{\tau}, \bar{t} \right\}$$

$$\mathbf{B}_{634} = \left\{ u, d, \nu_\mu, \mu, \bar{\nu}_\mu, s, \tau, \bar{\tau} \right\}$$

$$\mathbf{B}_{635} = \left\{ u, d, \nu_\mu, \mu, \bar{c}, \bar{s}, t, b \right\}$$

$$\mathbf{B}_{636} = \left\{ u, d, \nu_\mu, \bar{\nu}_\mu, \bar{\mu}, \bar{c}, \bar{\nu}_\tau, \bar{t} \right\}$$

$$\mathbf{B}_{637} = \left\{ u, d, \nu_\mu, \bar{\mu}, s, \bar{s}, \nu_\tau, \bar{b} \right\}$$

$$\mathbf{B}_{638} = \left\{ u, d, \nu_\mu, c, \nu_\tau, \tau, t, \bar{t} \right\}$$

$$\mathbf{B}_{639} = \left\{ u, d, \nu_\mu, c, \bar{\nu}_\tau, \bar{\tau}, b, \bar{b} \right\}$$

$$\mathbf{B}_{640} = \left\{ u, d, \mu, \bar{\nu}_\mu, c, \bar{c}, \nu_\tau, \bar{b} \right\}$$

$$\mathbf{B}_{641} = \left\{ u, d, \mu, \bar{\mu}, \nu_\tau, \bar{\tau}, b, \bar{t} \right\}$$

$$\mathbf{B}_{642} = \left\{ u, d, \mu, \bar{\mu}, \tau, \bar{\nu}_\tau, t, \bar{b} \right\}$$

$$\mathbf{B}_{643} = \left\{ u, d, \mu, c, s, \bar{s}, \bar{\nu}_\tau, \bar{t} \right\}$$

$$\mathbf{B}_{644} = \left\{ u, d, \bar{\nu}_\mu, \bar{\mu}, c, s, t, b \right\}$$

$$\mathbf{B}_{645} = \left\{ u, d, \bar{\nu}_\mu, \bar{s}, \nu_\tau, \bar{\nu}_\tau, \bar{\tau}, t \right\}$$

$$\mathbf{B}_{646} = \left\{ u, d, \bar{\nu}_\mu, \bar{s}, \tau, b, \bar{t}, \bar{b} \right\}$$

$$\mathbf{B}_{647} = \left\{ u, d, \bar{\mu}, c, \bar{c}, \bar{s}, \tau, \bar{\tau} \right\}$$

$$\mathbf{B}_{648} = \left\{ u, d, s, \bar{c}, \nu_\tau, \tau, \bar{\nu}_\tau, b \right\}$$

$$\mathbf{B}_{649} = \left\{ u, d, s, \bar{c}, \bar{\tau}, t, \bar{t}, \bar{b} \right\}$$

$$\mathbf{B}_{650} = \left\{ u, \bar{u}, \nu_\mu, \mu, \bar{\mu}, \bar{s}, \tau, \bar{t} \right\}$$

$$\mathbf{B}_{651} = \left\{ u, \bar{u}, \nu_\mu, \mu, c, s, t, \bar{b} \right\}$$

$$\mathbf{B}_{652} = \left\{ u, \bar{u}, \nu_\mu, \bar{\nu}_\mu, \nu_\tau, \bar{\tau}, \bar{t}, \bar{b} \right\}$$

$$\mathbf{B}_{653} = \left\{ u, \bar{u}, \nu_\mu, \bar{\nu}_\mu, \tau, \bar{\nu}_\tau, t, b \right\}$$

$$\mathbf{B}_{654} = \left\{ u, \bar{u}, \nu_\mu, \bar{\mu}, c, \bar{c}, \nu_\tau, b \right\}$$

$$\mathbf{B}_{655} = \left\{ u, \bar{u}, \nu_\mu, s, \bar{c}, \bar{s}, \bar{\nu}_\tau, \bar{\tau} \right\}$$

$$\mathbf{B}_{656} = \left\{ u, \bar{u}, \mu, \bar{\nu}_\mu, \bar{\mu}, c, \bar{\nu}_\tau, \bar{\tau} \right\}$$

$$\mathbf{B}_{657} = \left\{ u, \bar{u}, \mu, \bar{\nu}_\mu, s, \bar{s}, \nu_\tau, b \right\}$$

$$\mathbf{B}_{658} = \left\{ u, \bar{u}, \mu, \bar{c}, \nu_\tau, \tau, \bar{\tau}, t \right\}$$

$$\mathbf{B}_{659} = \left\{ u, \bar{u}, \mu, \bar{c}, \bar{\nu}_\tau, b, \bar{t}, \bar{b} \right\}$$

$$\mathbf{B}_{660} = \left\{ u, \bar{u}, \bar{\nu}_\mu, \bar{\mu}, \bar{c}, \bar{s}, t, \bar{b} \right\}$$

$$\mathbf{B}_{661} = \left\{ u, \bar{u}, \bar{\nu}_\mu, c, s, \bar{c}, \tau, \bar{t} \right\}$$

$$\mathbf{B}_{662} = \left\{ u, \bar{u}, \bar{\mu}, s, \nu_\tau, \bar{\nu}_\tau, t, \bar{t} \right\}$$

$$\mathbf{B}_{663} = \left\{ u, \bar{u}, \bar{\mu}, s, \tau, \bar{\tau}, b, \bar{b} \right\}$$

$$\mathbf{B}_{664} = \left\{ u, \bar{u}, c, \bar{s}, \nu_\tau, \tau, \bar{\nu}_\tau, \bar{b} \right\}$$

$$\mathbf{B}_{665} = \left\{ u, \bar{u}, c, \bar{s}, \bar{\tau}, t, b, \bar{t} \right\}$$

$$\mathbf{B}_{666} = \left\{ u, \bar{d}, \nu_\mu, \mu, \bar{\nu}_\mu, \bar{\mu}, \nu_\tau, t \right\}$$

$$\mathbf{B}_{667} = \left\{ u, \bar{d}, \nu_\mu, \mu, c, \bar{c}, \tau, \bar{\nu}_\tau \right\}$$

$$\mathbf{B}_{668} = \left\{ u, \bar{d}, \nu_\mu, \bar{\nu}_\mu, s, \bar{c}, b, \bar{b} \right\}$$

$$\mathbf{B}_{669} = \left\{ u, \bar{d}, \nu_\mu, \bar{\mu}, c, s, \bar{\tau}, \bar{t} \right\}$$

$$\mathbf{B}_{670} = \left\{ u, \bar{d}, \nu_\mu, \bar{s}, \nu_\tau, \bar{\nu}_\tau, b, \bar{t} \right\}$$

$$\mathbf{B}_{671} = \left\{ u, \bar{d}, \nu_\mu, \bar{s}, \tau, \bar{\tau}, t, \bar{b} \right\}$$

$$\mathbf{B}_{672} = \left\{ u, \bar{d}, \mu, \bar{\nu}_\mu, \bar{c}, \bar{s}, \bar{\tau}, \bar{t} \right\}$$

$$\mathbf{B}_{673} = \left\{ u, \bar{d}, \mu, \bar{\mu}, c, \bar{s}, b, \bar{b} \right\}$$

$$\mathbf{B}_{674} = \left\{ u, \bar{d}, \mu, s, \nu_\tau, \bar{\nu}_\tau, \bar{\tau}, \bar{b} \right\}$$

$$\mathbf{B}_{675} = \left\{ u, \bar{d}, \mu, s, \tau, t, b, \bar{t} \right\}$$

$$\mathbf{B}_{676} = \left\{ u, \bar{d}, \bar{\nu}_\mu, \bar{\mu}, s, \bar{s}, \tau, \bar{\nu}_\tau \right\}$$

$$\mathbf{B}_{677} = \left\{ u, \bar{d}, \bar{\nu}_\mu, c, \nu_\tau, \tau, \bar{\tau}, b \right\}$$

$$\mathbf{B}_{678} = \left\{ u, \bar{d}, \bar{\nu}_\mu, c, \bar{\nu}_\tau, t, \bar{t}, \bar{b} \right\}$$

$$\mathbf{B}_{679} = \left\{ u, \bar{d}, \bar{\mu}, \bar{c}, \nu_\tau, \tau, \bar{t}, \bar{b} \right\}$$

$$\mathbf{B}_{680} = \left\{ u, \bar{d}, \bar{\mu}, \bar{c}, \bar{\nu}_\tau, \bar{\tau}, t, b \right\}$$

$$\mathbf{B}_{681} = \left\{ u, \bar{d}, c, s, \bar{c}, \bar{s}, \nu_\tau, t \right\}$$

$$\mathbf{B}_{682} = \left\{ d, \bar{u}, \nu_\mu, \mu, \bar{\nu}_\mu, \bar{\mu}, b, \bar{b} \right\}$$

$$\mathbf{B}_{683} = \left\{ d, \bar{u}, \nu_\mu, \mu, c, \bar{c}, \bar{\tau}, \bar{t} \right\}$$

$$\mathbf{B}_{684} = \left\{ d, \bar{u}, \nu_\mu, \bar{\nu}_\mu, s, \bar{c}, \nu_\tau, t \right\}$$

$$\mathbf{B}_{685} = \left\{ d, \bar{u}, \nu_\mu, \bar{\mu}, c, s, \tau, \bar{\nu}_\tau \right\}$$

$$\mathbf{B_{686}} = \left\{ d, \bar{u}, \nu_\mu, \bar{s}, \nu_\tau, \tau, \bar{\tau}, b \right\}$$

$$\mathbf{B_{687}} = \left\{ d, \bar{u}, \nu_\mu, \bar{s}, \bar{\nu}_\tau, t, \bar{t}, \bar{b} \right\}$$

$$\mathbf{B_{688}} = \left\{ d, \bar{u}, \mu, \bar{\nu}_\mu, \bar{c}, \bar{s}, \tau, \bar{\nu}_\tau \right\}$$

$$\mathbf{B_{689}} = \left\{ d, \bar{u}, \mu, \bar{\mu}, c, \bar{s}, \nu_\tau, t \right\}$$

$$\mathbf{B_{690}} = \left\{ d, \bar{u}, \mu, s, \nu_\tau, \tau, \bar{t}, \bar{b} \right\}$$

$$\mathbf{B_{691}} = \left\{ d, \bar{u}, \mu, s, \bar{\nu}_\tau, \bar{\tau}, t, b \right\}$$

$$\mathbf{B_{692}} = \left\{ d, \bar{u}, \bar{\nu}_\mu, \bar{\mu}, s, \bar{s}, \bar{\tau}, \bar{t} \right\}$$

$$\mathbf{B_{693}} = \left\{ d, \bar{u}, \bar{\nu}_\mu, c, \nu_\tau, \bar{\nu}_\tau, b, \bar{t} \right\}$$

$$\mathbf{B_{694}} = \left\{ d, \bar{u}, \bar{\nu}_\mu, c, \tau, \bar{\tau}, t, \bar{b} \right\}$$

$$\mathbf{B_{695}} = \left\{ d, \bar{u}, \bar{\mu}, \bar{c}, \nu_\tau, \bar{\nu}_\tau, \bar{\tau}, \bar{b} \right\}$$

$$\mathbf{B_{696}} = \left\{ d, \bar{u}, \bar{\mu}, \bar{c}, \tau, t, b, \bar{t} \right\}$$

$$\mathbf{B_{697}} = \left\{ d, \bar{u}, c, s, \bar{c}, \bar{s}, b, \bar{b} \right\}$$

$$\mathbf{B_{698}} = \left\{ d, \bar{d}, \nu_\mu, \mu, \bar{\mu}, \bar{s}, \bar{\nu}_\tau, \bar{\tau} \right\}$$

$$\mathbf{B_{699}} = \left\{ d, \bar{d}, \nu_\mu, \mu, c, s, \nu_\tau, b \right\}$$

$$\mathbf{B}_{700} = \left\{ d, \bar{d}, \nu_\mu, \bar{\nu}_\mu, \nu_\tau, \tau, \bar{\nu}_\tau, \bar{b} \right\}$$

$$\mathbf{B}_{701} = \left\{ d, \bar{d}, \nu_\mu, \bar{\nu}_\mu, \bar{\tau}, t, b, \bar{t} \right\}$$

$$\mathbf{B}_{702} = \left\{ d, \bar{d}, \nu_\mu, \bar{\mu}, c, \bar{c}, t, \bar{b} \right\}$$

$$\mathbf{B}_{703} = \left\{ d, \bar{d}, \nu_\mu, s, \bar{c}, \bar{s}, \tau, \bar{t} \right\}$$

$$\mathbf{B}_{704} = \left\{ d, \bar{d}, \mu, \bar{\nu}_\mu, \bar{\mu}, c, \tau, \bar{t} \right\}$$

$$\mathbf{B}_{705} = \left\{ d, \bar{d}, \mu, \bar{\nu}_\mu, s, \bar{s}, t, \bar{b} \right\}$$

$$\mathbf{B}_{706} = \left\{ d, \bar{d}, \mu, \bar{c}, \nu_\tau, \bar{\nu}_\tau, t, \bar{t} \right\}$$

$$\mathbf{B}_{707} = \left\{ d, \bar{d}, \mu, \bar{c}, \tau, \bar{\tau}, b, \bar{b} \right\}$$

$$\mathbf{B}_{708} = \left\{ d, \bar{d}, \bar{\nu}_\mu, \bar{\mu}, \bar{c}, \bar{s}, \nu_\tau, b \right\}$$

$$\mathbf{B}_{709} = \left\{ d, \bar{d}, \bar{\nu}_\mu, c, s, \bar{c}, \bar{\nu}_\tau, \bar{\tau} \right\}$$

$$\mathbf{B}_{710} = \left\{ d, \bar{d}, \bar{\mu}, s, \nu_\tau, \tau, \bar{\tau}, t \right\}$$

$$\mathbf{B}_{711} = \left\{ d, \bar{d}, \bar{\mu}, s, \bar{\nu}_\tau, b, \bar{t}, \bar{b} \right\}$$

$$\mathbf{B}_{712} = \left\{ d, \bar{d}, c, \bar{s}, \nu_\tau, \bar{\tau}, \bar{t}, \bar{b} \right\}$$

$$\mathbf{B}_{713} = \left\{ d, \bar{d}, c, \bar{s}, \tau, \bar{\nu}_\tau, t, b \right\}$$

$$\mathbf{B}_{714} = \left\{ \bar{u}, \bar{d}, \nu_\mu, \mu, \bar{\nu}_\mu, s, \bar{\nu}_\tau, \bar{t} \right\}$$

$$\mathbf{B}_{715} = \left\{ \bar{u}, \bar{d}, \nu_\mu, \mu, \bar{c}, \bar{s}, \nu_\tau, \bar{b} \right\}$$

$$\mathbf{B}_{716} = \left\{ \bar{u}, \bar{d}, \nu_\mu, \bar{\nu}_\mu, \bar{\mu}, \bar{c}, \tau, \bar{\tau} \right\}$$

$$\mathbf{B}_{717} = \left\{ \bar{u}, \bar{d}, \nu_\mu, \bar{\mu}, s, \bar{s}, t, b \right\}$$

$$\mathbf{B}_{718} = \left\{ \bar{u}, \bar{d}, \nu_\mu, c, \nu_\tau, \bar{\nu}_\tau, \bar{\tau}, t \right\}$$

$$\mathbf{B}_{719} = \left\{ \bar{u}, \bar{d}, \nu_\mu, c, \tau, b, \bar{t}, \bar{b} \right\}$$

$$\mathbf{B}_{720} = \left\{ \bar{u}, \bar{d}, \mu, \bar{\nu}_\mu, c, \bar{c}, t, b \right\}$$

$$\mathbf{B}_{721} = \left\{ \bar{u}, \bar{d}, \mu, \bar{\mu}, \nu_\tau, \tau, \bar{\nu}_\tau, b \right\}$$

$$\mathbf{B}_{722} = \left\{ \bar{u}, \bar{d}, \mu, \bar{\mu}, \bar{\tau}, t, \bar{t}, \bar{b} \right\}$$

$$\mathbf{B}_{723} = \left\{ \bar{u}, \bar{d}, \mu, c, s, \bar{s}, \tau, \bar{\tau} \right\}$$

$$\mathbf{B}_{724} = \left\{ \bar{u}, \bar{d}, \bar{\nu}_\mu, \bar{\mu}, c, s, \nu_\tau, \bar{b} \right\}$$

$$\mathbf{B}_{725} = \left\{ \bar{u}, \bar{d}, \bar{\nu}_\mu, \bar{s}, \nu_\tau, \tau, t, \bar{t} \right\}$$

$$\mathbf{B}_{726} = \left\{ \bar{u}, \bar{d}, \bar{\nu}_\mu, \bar{s}, \bar{\nu}_\tau, \bar{\tau}, b, \bar{b} \right\}$$

$$\mathbf{B}_{727} = \left\{ \bar{u}, \bar{d}, \bar{\mu}, c, \bar{c}, \bar{s}, \bar{\nu}_\tau, \bar{t} \right\}$$

$$\mathbf{B}_{728} = \left\{ \bar{u}, \bar{d}, s, \bar{c}, \nu_\tau, \bar{\tau}, b, \bar{t} \right\}$$

$$\mathbf{B}_{729} = \left\{ \bar{u}, \bar{d}, s, \bar{c}, \tau, \bar{\nu}_\tau, t, \bar{b} \right\}$$

$$\mathbf{B}_{730} = \left\{ \nu_\mu, \mu, \bar{\nu}_\mu, \bar{\mu}, c, s, \bar{c}, \bar{s} \right\}$$

$$\mathbf{B}_{731} = \left\{ \nu_\mu, \mu, \bar{\nu}_\mu, \bar{c}, \nu_\tau, \bar{\nu}_\tau, \bar{\tau}, b \right\}$$

$$\mathbf{B}_{732} = \left\{ \nu_\mu, \mu, \bar{\nu}_\mu, \bar{c}, \tau, t, \bar{t}, \bar{b} \right\}$$

$$\mathbf{B}_{733} = \left\{ \nu_\mu, \mu, \bar{\mu}, c, \nu_\tau, \bar{\nu}_\tau, \bar{t}, \bar{b} \right\}$$

$$\mathbf{B}_{734} = \left\{ \nu_\mu, \mu, \bar{\mu}, c, \tau, \bar{\tau}, t, b \right\}$$

$$\mathbf{B}_{735} = \left\{ \nu_\mu, \mu, s, \bar{s}, \nu_\tau, \tau, \bar{\nu}_\tau, t \right\}$$

$$\mathbf{B}_{736} = \left\{ \nu_\mu, \mu, s, \bar{s}, \bar{\tau}, b, \bar{t}, \bar{b} \right\}$$

$$\mathbf{B}_{737} = \left\{ \nu_\mu, \bar{\nu}_\mu, \bar{\mu}, s, \nu_\tau, \tau, b, \bar{t} \right\}$$

$$\mathbf{B}_{738} = \left\{ \nu_\mu, \bar{\nu}_\mu, \bar{\mu}, s, \bar{\nu}_\tau, \bar{\tau}, t, \bar{b} \right\}$$

$$\mathbf{B}_{739} = \left\{ \nu_\mu, \bar{\nu}_\mu, c, \bar{s}, \nu_\tau, t, b, \bar{b} \right\}$$

$$\mathbf{B}_{740} = \left\{ \nu_\mu, \bar{\nu}_\mu, c, \bar{s}, \tau, \bar{\nu}_\tau, \bar{\tau}, \bar{t} \right\}$$

$$\mathbf{B}_{741} = \left\{ \nu_\mu, \bar{\mu}, \bar{c}, \bar{s}, \nu_\tau, \bar{\tau}, t, \bar{t} \right\}$$

$$\mathbf{B}_{742} = \left\{ \nu_\mu, \bar{\mu}, \bar{c}, \bar{s}, \tau, \bar{\nu}_\tau, b, \bar{b} \right\}$$

$$\mathbf{B}_{743} = \left\{ \nu_\mu, c, s, \bar{c}, \nu_\tau, \tau, \bar{\tau}, \bar{b} \right\}$$

$$\mathbf{B}_{744} = \left\{ \nu_\mu, c, s, \bar{c}, \bar{\nu}_\tau, t, b, \bar{t} \right\}$$

$$\mathbf{B}_{745} = \left\{ \mu, \bar{\nu}_\mu, \bar{\mu}, \bar{s}, \nu_\tau, \tau, \bar{\tau}, \bar{b} \right\}$$

$$\mathbf{B}_{746} = \left\{ \mu, \bar{\nu}_\mu, \bar{\mu}, \bar{s}, \bar{\nu}_\tau, t, b, \bar{t} \right\}$$

$$\mathbf{B}_{747} = \left\{ \mu, \bar{\nu}_\mu, c, s, \nu_\tau, \bar{\tau}, t, \bar{t} \right\}$$

$$\mathbf{B}_{748} = \left\{ \mu, \bar{\nu}_\mu, c, s, \tau, \bar{\nu}_\tau, b, \bar{b} \right\}$$

$$\mathbf{B}_{749} = \left\{ \mu, \bar{\mu}, s, \bar{c}, \nu_\tau, t, b, \bar{b} \right\}$$

$$\mathbf{B}_{750} = \left\{ \mu, \bar{\mu}, s, \bar{c}, \tau, \bar{\nu}_\tau, \bar{\tau}, \bar{t} \right\}$$

$$\mathbf{B}_{751} = \left\{ \mu, c, \bar{c}, \bar{s}, \nu_\tau, \tau, b, \bar{t} \right\}$$

$$\mathbf{B}_{752} = \left\{ \mu, c, \bar{c}, \bar{s}, \bar{\nu}_\tau, \bar{\tau}, t, \bar{b} \right\}$$

$$\mathbf{B}_{753} = \left\{ \bar{\nu}_\mu, \bar{\mu}, c, \bar{c}, \nu_\tau, \tau, \bar{\nu}_\tau, t \right\}$$

$$\mathbf{B}_{754} = \left\{ \bar{\nu}_\mu, \bar{\mu}, c, \bar{c}, \bar{\tau}, b, \bar{t}, \bar{b} \right\}$$

$$\mathbf{B}_{755} = \left\{ \bar{\nu}_\mu, s, \bar{c}, \bar{s}, \nu_\tau, \bar{\nu}_\tau, \bar{t}, \bar{b} \right\}$$

$$\mathbf{B}_{756} = \left\{ \bar{\nu}_\mu, s, \bar{c}, \bar{s}, \tau, \bar{\tau}, t, b \right\}$$

$$\mathbf{B}_{757} = \left\{ \bar{\mu}, c, s, \bar{s}, \nu_\tau, \bar{\nu}_\tau, \bar{\tau}, b \right\}$$

$$\mathbf{B}_{758} = \left\{ \bar{\mu}, c, s, \bar{s}, \tau, t, \bar{t}, \bar{b} \right\}$$

$$\mathbf{B}_{759} = \left\{ \nu_\tau, \tau, \bar{\nu}_\tau, \bar{\tau}, t, b, \bar{t}, \bar{b} \right\}$$

References

[1] Ashay Dharwadker, *A New Proof of the Four Colour Theorem*, **http://www.dharwadker.org/** , 2000.

[2] Ashay Dharwadker, *Grand Unification of the Standard Model with Quantum Gravity*, **http://www.dharwadker.org/standard_model/** , 2008.

[3] Ashay Dharwadker and Vladimir Khachatryan, *Higgs Boson Mass predicted by the Four Color Theorem*, **http://arxiv.org/abs/0912.5189** , 2009.

[4] Albert Einstein, *On the Generalized Theory of Gravitation*, Scientific American, 1950.

[5] Albert Einstein, *The Meaning of Relativity*, Princeton University Press, 1956.

[6] Hermann Minkowski, *Raum und Zeit*, 1908.

[7] Richard P. Feynman, *The Special Theory of Relativity*, The Feynman Lectures on Physics, Vol. 1, Addison-Wesley, 1965.

[8] Max Plank, *Über irreversible Strahlungsvorgänge*, 1899.

[9] Paul A.M. Dirac, *The Quantum Theory of the Electron*, Proc. Roy. Soc. A 117, 1928.

[10] Paul A.M. Dirac, *Quantised Singularities in the Electromagnetic Field*, Proc. Roy. Soc. A 133, 1931.

[11] Gerard 't Hooft, *The Topological Mechanism for Permanent Quark Confinement in a Non-Abelian Gauge Theory*, Physica Scripta, Vol.25, 1980.

[12] Peter Higgs, *Broken Symmetries and the Masses of Gauge Bosons*, Phys. Rev. Lett. 13, 1964.

[13] Leonard I. Schiff, *Quantum Mechanics*, McGraw-Hill Inc, 1968.

[14] Haskell Curry, *Foundations of Mathematical Logic*, Dover, 1963.

[15] Hideki Yukawa, *On the Interaction of Elementary Particles*, Proc. Phys. Math. Soc. Japan, 17, 1935.

[16] Thomas W.B. Kibble, *Global conservation laws and massless particles*, Phys. Rev. Lett. 13, 1964.

[17] Eugene P. Wigner, *On unitary representations of the inhomogeneous Lorentz group*, Annals of Mathematics 40, 1939.

[18] Wolfgang Pauli, *The Connection Between Spin and Statistics*, Phys. Rev. 58, 1940.

[19] Hendrik B.G. Casimir, *Rotation of a Rigid Body in Quantum Mechanics*, Dissertation, Leiden, 1931.

[20] Emmy Noether, *Hyperkomplexe Grössen und Darstellungstheorie*, Math. Zeit. 30, 1929.

[21] Émile Mathieu, *Sur la fonction cinq fois transitive de 24 quantités*, Liouville Journal XVIII, 1873.

[22] Ernst Witt, *Die 5-fach transitiven Gruppen von Mathieu*, Abh. Math. Sem. Univ. Hamburg 12, 1938.

[23] Ashay Dharwadker, *The Witt Design*, **http://www.dharwadker.org/witt.html** , 2001.

[24] Particle Data Group, *Particle Listings*, **http://pdg.lbl.gov/** , 2010.

www.ingramcontent.com/pod-product-compliance
Lightning Source LLC
Chambersburg PA
CBHW050730180526
45159CB00003B/1178
* 9 7 8 1 4 6 6 4 0 3 9 2 5 *